OFFSHORE MOORINGS

Proceedings of a conference organized by the Institution of
Civil Engineers and held in East Kilbride on 10 March 1982

THOMAS TELFORD LTD, LONDON, 1982

Published for the Institution of Civil Engineers by Thomas Telford Limited, PO Box 101, 26–34 Old Street, London EC1P 1JH

First published 1982

Organizing Committee: D. G. Owen, S. M. Abdi, G. Elliott, T. Ridley, E. A. Ross and R. G. Standing

Offshore moorings.
   1. Oil well drilling, submarine — Congresses
   2. Oil well drilling rigs — Congresses
   3. Offshore structures — Congresses
   4. Deep-sea moorings — Congresses
   627' 98   TN863

ISBN: 0 7277 0158 4

Printed in Great Britain by The Thetford Press Limited, Thetford, Norfolk

Contents

1 Environmental loading and response

E. C. BOWERS, PhD, DIC, Hydraulics Research Station and
R. G. STANDING, PhD, National Maritime Institute

SYNOPSIS. Environmental forces on offshore structures are
caused by a combination of wind, waves and current. Of
these three components, wave forces are the most complex.
Separate effects of wind, waves and current are discussed,
as well as coupling between waves and currents and the combined effect of all three on moored structures. Various
typical motions of moored structures and their moorings are
used as a means of introducing the environmental forces,
and comments are made on the ability of physical and mathematical models to describe the physics involved. Tethered
buoyant platforms and single point moorings are used to
illustrate the various types of response and associated
design problems.

INTRODUCTION

1. The costs of conventional fixed oil-production platforms rise rapidly as the water depth increases. They also
become more compliant; their primary response frequencies
enter the wave spectrum, and fatigue problems arise. A number of very compliant structures are now being considered
for deep-water applications. These include the tethered
buoyant platform, now nearing reality in the Hutton field,
and the guyed tower. Various other compliant systems will
be required for the associated activities of exploration,
off-loading and servicing: for example, ships and semisubmersibles at multi-point or single-point moorings, or with
dynamic positioning systems. Wave energy devices pose similar mooring problems. In all cases the system is made compliant so that it responds to wave loading rather than
resisting those forces. In doing so, however, new problems,
like low frequency responses, arise.

2. Such systems can be represented as arrays of intercon-

nected springs and masses with certain modes of natural response. Designers usually try to ensure that any natural frequencies lie well outside the range of the wave spectrum. This policy does not, however, guarantee immunity from dynamic response. Various higher-order wave forces and other effects, such as vortex-shedding, can excite both high- and low-frequency response. These non-linear processes are difficult to predict theoretically, and pose physical modelling and scaling difficulties. There has been a recent upsurge of research interest in this general area. Problems of vortex-shedding and high-frequency response are mainly of local interest, for the design of risers, tethers and individual members. Low-frequency motions, however, affect the whole system. Typical of these are the surge, sway and yaw motions of a tethered buoyant platform, with periods of about a minute, similar motions of moored vessels in harbours, and the 'fishtailing' of ships at single point moorings Problems relating to these particular systems are discussed in greater detail later in this paper.

3. It is generally accepted that, to date, the oil industry has applied large safety factors in their mooring designs. Now, with the requirement to go to ever deeper water, mooring costs increase significantly and the need arises for safe but economic moorings. There is a similar need in the wave energy programme where every extra fraction of a penny per kilowatt of electricity produced, will count. With the need for economic moorings comes the requirement for an accurate description of the motions of moored structures.

4. There are two different descriptions of wave forces. One method, which is widely used in the design of offshore structures employs the idea of a design wave and this is called deterministic. The other method, called spectral, uses the idea of a superposition of a number of wave components of random phase relative to one another with a spectrum of wave energy. This approach results in a statistical description of wave loading.

5. Both methods can allow to some degree for the non-linear boundary condition that is satisfied by wave motion at the free surface. In the deterministic approach a high order wave theory is normally used to describe a unidirectional single period wave of large amplitude. In the spectral method a Stokes expansion of the basic wave equations can be used to obtain corrections to the superposition of linear wave components. Second order terms in this expansion, proportional to the square of wave amplitude, can cause a significant proportion of the load in moorings for offshore

structures (see paragraph 12). In third order, interactions between the wave components take place resulting in a transfer of energy to waves at different periods. These interactions are important in determining the shape of the wave spectrum. In a comparison (1) between measurements of orbital wave velocities in a storm and predictions from various wave theories it was found that a linear spectral description which allowed for a spread of wave energy in direction as well as in period gave better agreement than higher order theories for unidirectional waves.

6. The approaches used to calculate maximum forces and responses in the deterministic and spectral methods differ in the following way. In normal wave prediction techniques the extreme sea condition of interest will arise as one in which average parameters, like significant wave height and zero crossing wave period, remain steady over a specified duration. If not already defined it is normally possible to relate a wave spectrum to this sea state. A Pierson Moskowitz spectrum is used if the fetch involved is long enough to produce a fully developed sea and a JONSWAP spectrum can be used for shorter fetches. In the deterministic approach a design wave with a certain probability of occurrence over the specified duration can be defined in the given sea state and used to calculate maximum forces and responses. The period of the design wave can be expected to be close to the period at which the peak occurs in the wave spectrum. In the spectral approach, forces and responses are worked out first as functions of the wave parameters. In many cases of interest these functions are non-linear and it is necessary to study the probability distributions of such functions to obtain maximum forces and responses. For structures with resonant responses a less extreme sea state may excite a larger response if it has more energy at resonant periods. Therefore, care is required in the selection of design conditions. The spectral approach lends itself particularly to the study of the statistics of forces and responses in more frequently occurring sea states. Such considerations are useful for fatigue analysis. Many of the studies necessary to obtain the probability distributions of forces and responses and their maxima are only now being carried out and although the approach is more complex than that required in the deterministic method, it can be expected to yield more accurate results. Wave groups of various lengths arise naturally in the spectral description with various probabilities of occurrence and they are known to be important in the low-frequency structure response.

3

The idea of design groups is sometimes used in an effort to make such groupings deterministic. However, taken out of context in this way the probabilities of occurrence of such groups and the structure response that results, are ill-defined.

TYPES OF RESPONSE
Wave Period Response

7. Morison's equation. When the typical dimensions D of a structural member are small compared with the wavelength λ there will be little scattering of the wave field, and the force on the structure can be calculated using Morison's equation:

$$F = \rho C_m A\dot{u} + \tfrac{1}{2}\rho C_D Du|u|. \tag{1}$$

This formula is usually adequate when D/λ is less than about 0.2. F represents the force per unit length of the member, A its cross-sectional area and D its width. The first term is the inertial force, in phase with fluid acceleration $\dot{u}$. The second term is the drag force, associated with flow separation and depending on fluid velocity u. C_m and C_D are the inertia and drag coefficients. The relative importance of these two force components depends on the Keulegan-Carpenter number $K = u_m T/D$, where u_m is the maximum value of u during the wave cycle, and T is the wave period. The drag term dominates when $K \gtrsim 15$ (ie for slender tubular members just below the free surface in high waves), and the inertia term when $K < 15$.

8. The coefficients C_m, C_d depend on a range of parameters including the shape of the member, its roughness, K and the Reynolds number $R_e = u_m D/\nu$, where ν is the kinematic viscosity coefficient for water. Physical models accurately represent the forces in the Morison equation provided the Keulegan-Carpenter number is low. Models that use a spectrum with wave energy spread over both direction and period will be particularly realistic. At larger values of K, however, it is usually necessary for the model to be large enough for the flow regime to be supercritical (so that the drag coefficient becomes independent of Reynolds number), in order to represent both the drag and inertial forces accurately. It is difficult to achieve high Reynolds numbers in laboratory wave conditions and, therefore, difficult to obtain valid full-scale design data. Experiments on smooth circular columns at sea (Christchurch Bay and Ocean Test Structure) and in idealised planar flow (2) have achieved high values of R_e, and the results show similar average trends. Figure 1 shows how C_m, C_d and an overall force coefficient

4

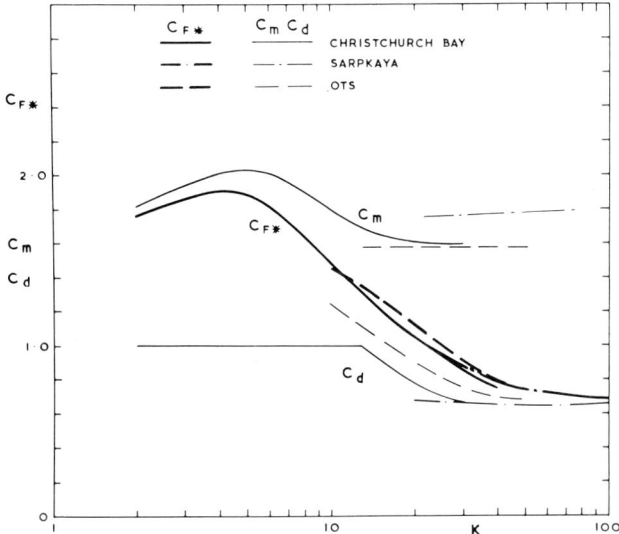

Fig 1. Force coefficients for smooth cylinders: variation
with K. Mean experimental results at large Re.

(3) vary with K. Superimposed on these mean trends there
is a considerable wave-to-wave variation in estimates of
C_m and C_d, associated with the shedding and convection of
individual vortices. Vortex-shedding also gives rise to a
lift force, at right angles to the direction of flow,
which is particularly significant in the range $5 < K < 30$.
Vortex-shedding can sometimes excite dynamic response of
flexible members, tethers and rises, with consequent fatigue
damage. Effects of vortex-shedding on total wave loads for
multi-member structures, however, are likely to be small.
Marine growth and roughness can increase wave loads very
substantially. In some cases the forces on roughened cylinders are double those on smooth members, and there is evidence for even larger increases with seaweed type growth (4).
    9. <u>Wave diffraction</u>. For ratios D/λ greater than about
0.2 the diffraction of waves by the structure becomes important. Physical limitations on wave steepness in such conditions keep the Keulegan-Carpenter number K small, so that
drag and effects of flow separation are generally unimpor-
tant. The effects of wave scattering and reflection are well
represented in physical models and they can also be calculated using computer programs, based on linear wave diffraction theory and assuming ideal potential flow (3). Computer

5

predictions of wave loads, response motions and tether forces often agree strikingly well with experimental data. Figure 2 is one such example.

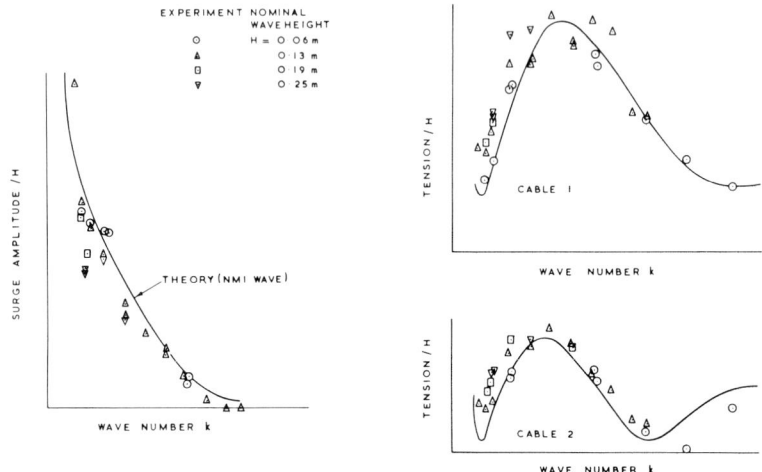

Fig 2. Comparison between theory and experiment for a TBP: surge amplitude and mooring cable tensions.

10. Response. To obtain the correct response of large floating moored structures in physical and computational models it is important that damping of the motion is correctly represented. Damping is normally due to a mixture of wave-making and viscous effects; the latter usually being an energy loss due to flow separation from the body. Viscous damping forces can sometimes become important locally: at the natural frequency, or at frequencies where inertial and buoyancy-type wave forces cancel. This occurs when the wave radiation damping is small, as it is for ships in roll, and for semisubmersible-type structures at the wave force cancellation frequency (5). Damping due to wave radiation or wave-making is, like diffraction, well represented in physical and computational models, and at reasonably large model scales viscous damping is also thought to be well represented in physical models, with flows being tubulent around structures with sharp boundaries. For example no serious scaling problems were apparent in a comparison made between the movements (which included low-frequency components) of a model vessel moored in a 1 to 100 scale model of a harbour and vessel movements in the real harbour, when the wave climate was correctly represented (6).

Low-frequency response

11. Hydrodynamic damping of slow oscillations of large moored structures (at periods of a minute or more) is normally low. As a result quite small low-frequency forces are able to build up significant oscillations, particularly in surge, sway and yaw. Figure 3 illustrates the process.

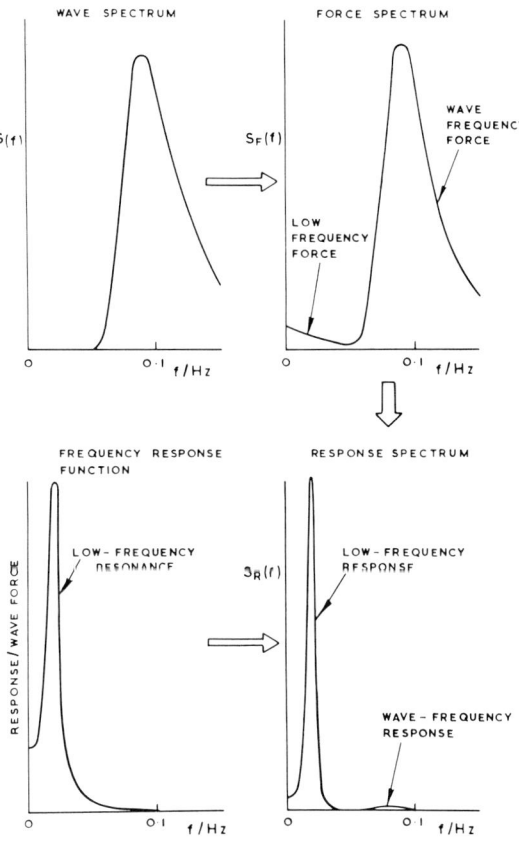

Fig 3. Relationship between wave force and response spectra for typical moored structure.

First-order waves (top left of diagram) give rise to both wave- and low-frequency forces (top right). With low hydrodynamic damping the system's response characteristic (bottom left) will be such as to attenuate motions at wave frequencies, but to amplify (relatively) the natural response component. A typical response spectrum is shown at bottom right of the figure. The mooring force depends on the response and can be many times the initial exciting force: their ratio is $\sqrt{2}\beta$ in regular waves at the natural frequency, where β is the fraction of critical damping present. The root mean square mooring force in an irregular sea is less sensitive to damping, depending on $\beta^{-\frac{1}{2}}$ only (7).

12. <u>Second order wave forces</u>. These forces have a mean component, which causes a static offset of the structure, as well as a slowly varying component which causes a low frequency response. Pinkster (8) has identified a number of contributions and they are listed as I to VI in figure 4

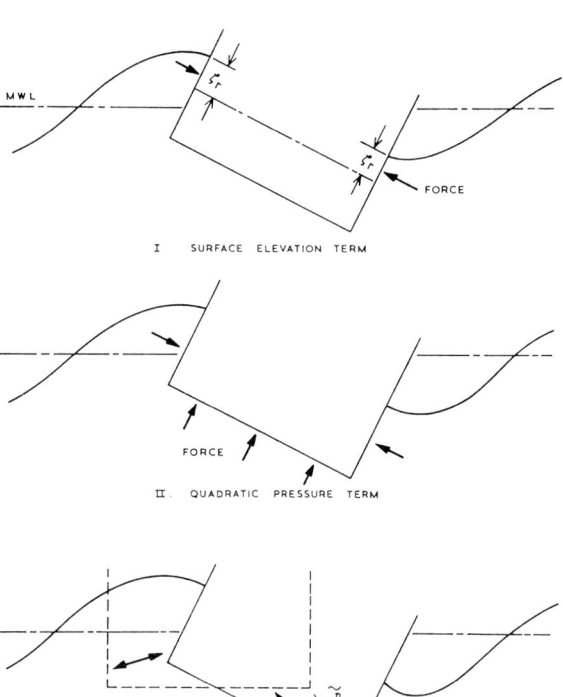

I   SURFACE ELEVATION TERM

II   QUADRATIC PRESSURE TERM

III   STRUCTURE DISPLACEMENT TERM

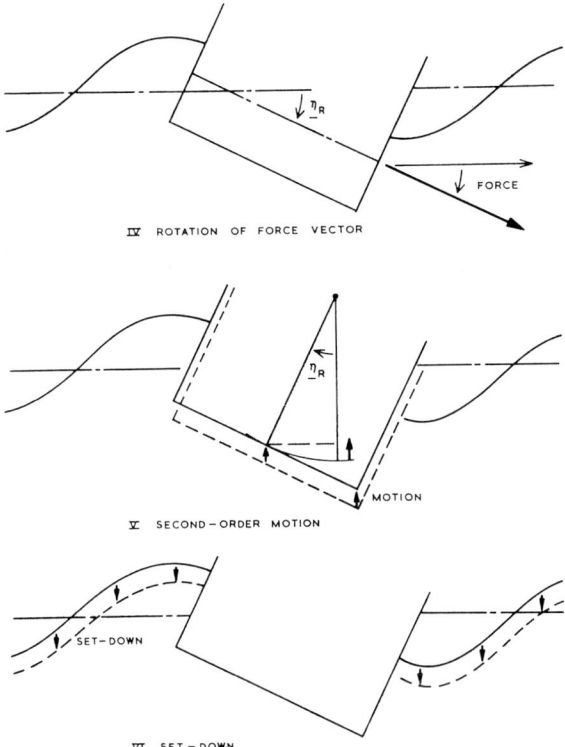

Fig 4. Six contributions to the second-order wave force.

for a structure in a regular wave. The first term is the
integral of the first order wave pressure over the area
between the mean waterline on the structure and the surface elevation. The second term is the integral of the
quadratic term in velocity in Bernoulli's equation over the
mean submerged area of the structure. The third term arises
from the second order part of the integral over the mean
submerged area of the first order wave pressure when that
pressure is evaluated at the position of the displaced body.
The fourth term arises from the rotation due to roll, pitch
and yaw of the total first order fluid force on the structure including the hydrostatic restoring force. The fifth
term ignored by Pinkster (8) but identified by Standing (7),
is a buoyancy force due to second order effects, and the
sixth term is the pressure in the second order wave field

acting over the mean submerged area. In regular waves these effects lead to a steady wave drift force. In an irregular sea, however, the same effects naturally lead to an additional slowly varying drift force due to the varying wave envelope. In particular, an important contribution to the effect (VI) of the second order wave field on the structure is made by set-down beneath wave groups (9). The spectral approach lends itself to a description of the slowly varying drift force which has a spectrum with frequencies that are differences between frequencies in the first order wave spectrum. If the structure causes significant diffraction of the primary waves then effects I to IV dominate the horizontal components of the drift force. This can be expected in short to moderate period seas where, typically, offshore structures are large enough to cause significant scattering. But in more extreme sea states, the wave periods are longer and structures will tend to move more with the waves thereby causing less diffraction. Effects I to V are then reduced and effect VI, due to the second order wave field, may make a significant contribution to the total horizontal slowly varying drift force on some structures.

13. This point is illustrated in figures 5 and 6 which

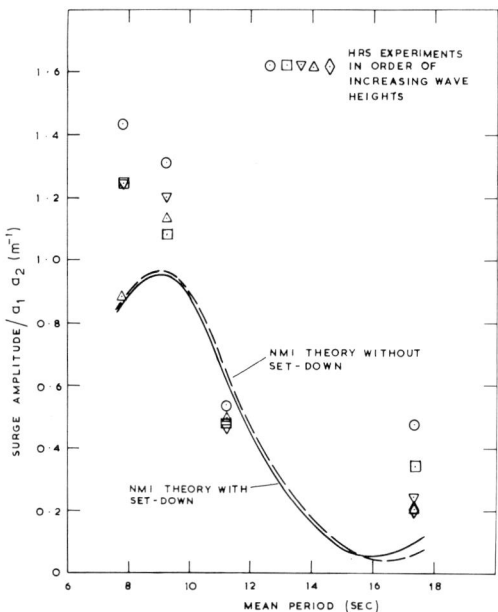

Fig 5. Second-order surge response operator for Oscillating Water Column.

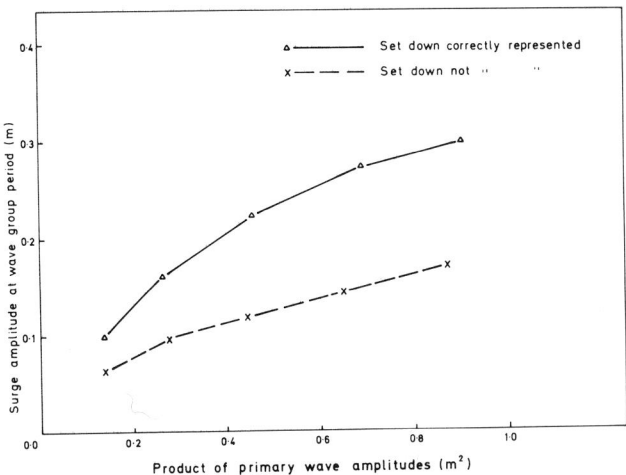

Fig 6.   Second-order surge of Oscillating Water Column
(mean wave = 17s).

show computational and model results for the slowly varying
surge response of a wave energy device, an oscillating
water column moored across the waves in a water depth of
60m. Calculations and experiments were both performed in
the simplest sea state which gives rise to low-frequency
second-order forcing: in regular wave groups, obtained by
superimposing two regular wave trains with slightly diffe-
rent frequencies σ_1, σ_2 and amplitudes a_1, a_2. The theoretical results shown in figure 5 were obtained using a computational model developed at NMI (7) (see also paragraph
14). The theoretical results obtained with the set-down
term diverge from those without the set-down term at long
mean wave periods only. It is clear, therefore, that
effects I to IV dominate the slow drift force for mean wave
periods less than about 15 seconds. The experimental results shown in both figures are taken from Ref 10; those
shown in figure 6 are for a mean wave period of about 17
seconds and the values marked by Δ are ones for which the
movement of the wave-maker was modified to produce the correct set-down beneath wave groups in the second order wave
field. In the absence of this modification additional
second order waves at the wave group period are produced
with an amplitude that is similar to that of set-down, and
their effect can be gauged by comparing results marked by
X with results marked by Δ in figure 6. Further, since the
amplitudes of free wave and set-down are similar, the relative importance of set-down can be gauged by the difference

between the two sets of experimental results. Clearly, for this mean wave period set-down is important. Similar tests carried out with shorter mean periods produced a slowly varying response that was relatively insensitive to whether or not set-down was correctly represented. These experimental results are qualitatively consistent with the theoretical results shown in figure 5. Although figure 5 shows that the second order surge, divided by the product of wave amplitudes, is lower for longer mean wave periods, the absolute amount of surge can be higher because longer period waves can reach greater heights before breaking.

14. Computer models that calculate second order wave forces are under development at present. It is convenient to group these forces under two main headings. Firstly, there are forces that can be calculated from products of first order quantities and they can be accurately calculated once the first order diffraction problem has been solved. Effects I to V in figure 4 come under this heading. Secondly, there are forces that require solution of the diffraction problem to second order and effect VI in figure 4 is of this type. Due to the complexity of solving second order diffraction problems approximate methods of calculating effect VI are used at present. NMI (7) for example, following Bowers, (11) simply integrate undisturbed second-order pressures over the vessel surface. Pinkster (8) includes one disturbance component, but neglects several others which may be equally important. The fundamental second-order wave is set-down beneath wave groups, but it is clear from the experimental results shown in figure 6 that the movement of wave-makers in physical models require modification to minimise the generation of additional spurious free waves at wave group periods. Work along these lines is now in progress at some laboratories (12, 13). The other non-linear wave forces (effects I to V in figure 4) are well represented in random wave physical models.

15. Two superimposed regular waves with frequencies σ_1, σ_2 give use to a second-order force with a non-zero mean and two oscillating components with frequencies $(\sigma_1+\sigma_2)$ and $(\sigma_1-\sigma_2)$. The difference-frequency component varies with the wave surface envelope, as shown in figure 7, but with a constant phase difference α_{12}. An irregular wave with spectral density $S(\sigma)$ similarly gives rise to mean and slowly-varying second-order forces. Pinkster (8) has shown that the mean force is

$$F = 2 \int_0^\infty S(\sigma) \ T(\sigma,\sigma) d\sigma$$

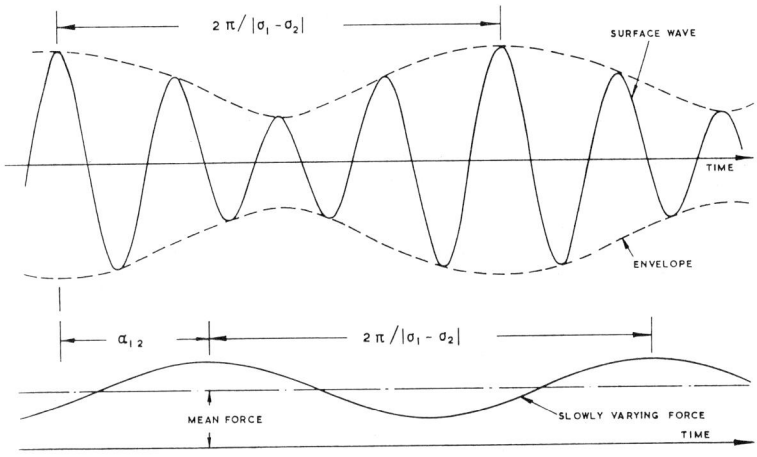

Fig 7. Surface wave and second-order drift force in a regular wave group: wave frequencies σ_1, σ_2.

and the slowly-varying force spectrum is

$$S_F(\sigma') = 8 \int_0^\infty S(\sigma+\sigma')S(\sigma)T(\sigma,\sigma')d\sigma .$$

Pinkster (8) and NMI (7) have developed computer programs, based on linear diffraction theory, which first evaluate the transfer function $T(\sigma,\sigma')$ for frequency pairs σ,σ', then evaluate the mean and low-frequency spectra of forcing and response. These programs have been validated experimentally, and fair agreement found, bearing in mind the difficulties and inaccuracies of both the experiments and numerical procedures.

16. A more approximate method, due to Newman (14) and Pinkster (8), is widely used in offshore design. It assumes that any low frequencies of interest are very much less than fundamental wave frequencies, and then replaces the difference-frequency function $T(\sigma,\sigma')$ above by the mean force component at the mean frequency $\sigma+\frac{1}{2}\sigma'$. The low-frequency force spectrum may then be estimated using information about mean forces in regular waves only. Mean forces are often more readily available to the designer than low-frequency components, either from model tests or from theory (15). Newman's method includes effects of wave diffraction, but neglects terms associated with spatial gradients of forcing such as set-down. Bowers (11) made largely complementary assumptions: he included gradient terms, but neglected wave

13

diffraction and effects of the vessel's response. Ref 7 asseses both of these approximate methods in relation to the more accurate numerical procedure described above. Newman's method was found to work well for ships in both surge and sway provided the natural response frequency is low enough. At higher frequencies neither approximate method is fully viable, and the more complete solution is preferred.

17. Ref 7 relates the effects of wave diffraction and gradient terms, in particular that due to set-down, to three simple parameters. The resulting force regimes agree qualitatively with the above conclusions about the Newman and Bowers procedures, with results discussed in paragraph 13 above, and with published data on ships, semisubmersibles and other structures. Diffraction terms tend to dominate when the natural frequency of the system is low, and the fundamental wave frequencies high; and vice-versa for the gradient terms. Semisubmersibles may be affected by both gradient and diffraction effects, also by drag (16). Ref 15 suggests that drag will have little effect on the drift force provided $H\lambda^2/D^3 \lesssim 60$.

18. <u>Non-linear moorings</u>. These can also cause slowly-varying responses. Such motions take place even in regular waves at a subharmonic of the wave period. The subharmonics that occur are normally the ones that are nearest to the resonant periods of the structure on its moorings. Lean (17) has described such motions for vessels swaying against fenders. The non-linearity here is caused by the fenders being stiffer than mooring lines. Mathieu instabilities are thought possible in the motions of Tethered Buoyant Platforms, where the horizontal restoring force from moorings has a harmonic component (see paragraph 32). Catenary moorings may also give rise to subharmonic responses. Damping, which may be non-linear as well, is particularly important in determining which subharmonics can occur, and in practice it is likely that only second or third subharmonics of the wave period will be present.

19. From paragraphs 11 to 18 it is clear that, in general, it is necessary to allow for non-linearity in the wave force, in the moorings and in the damping of structure responses to obtain an accurate description of slowly varying responses. By working in frequency space the computer models described in paragraphs 12-15 can be used to obtain wave period and low frequency responses of structures with linear damping and linear moorings. In general, though, it is necessary to work in the time domain to represent the full non-linea-

rity of moored structures. Simulation models that attempt
this step are now under development but one requirement,
that applies equally to physical models, is that the length
of the cycle of random wave activity be long enough to allow
the frequencies of occurrence of various patterns of wave
grouping to establish themselves. For example, the results
shown in table 1 were obtained at HRS for a model of a
moored structure, with a resonant surge period of about a
minute, subjected to varying lengths of repeating cycles
of random wave activity, all of which had the same wave
spectrum. In these examples the significant surge movement is defined as four times the standard deviation. The
very long cycle lasting the equivalent of 44 hours was not
subject to spectral analysis and so the amount of surge at
the wave period is not shown. But it can be estimated from
the average of the four similar values obtained using shorter cycles. This gives 3.55m and subtracting the square of
this value from the square of the total surge movement and
then taking the square root of the resultant gives an estimate of 2.93m for the low-frequency significant surge movement at the natural period. Comparing this value with values

Table 1. Effect of varying length of psuedo-random wave
cycle

Length of psuedo-random cycle (full scale)	2½mins	10½mins	21mins	42mins	44hrs 27
No of natural surge periods in cycle	2½	10	20	40 ·	2581
Significant wave height (m)	5.1	5.0	5.0	4.9	5.0
Significant surge movement (m) at wave period	3.6	3.6	3.5	3.5	
Significant surge movement (m) at natural period	3.7	1.6	3.3	2.9	
Total significant surge (m)	5.1	3.9	4.8	4.6	4.6

for the same quantity obtained from shorter cycles shows that a cycle of random waves lasting at least 40 natural surge periods is required to obtain a stable low-frequency response. In particular, the cycle lasting 10 natural periods gave only about half the true low-frequency response. For structures with very low damping of their resonant motions even longer lengths of random wave simulation may be required to give a stable estimate of the low-frequency response (18).

20. Reduction of mooring loads by passive means. It may be possible to reduce the second order wave force by making use of the following effect. Since the mean water level is set-down beneath groups of large waves, an increase in radiation pressure due to increasing wave height in a wave group is accompanied by a reduction in water pressure due to set-down. Thus, by making the submerged surface area of the structure large enough to collect the second order wave pressure due to set-down it may be possible to counter-act the force acting at the waterline due to radiation stress which is one of the dominant contributions to the slowly varying force. Such a reduction was demonstrated recently in experiments carried out on a model of a wave energy device (10). In this case a plate was added that extended almost the full water depth beneath the device. However, in some structures, like semisubmersibles, the area presented to the radiation pressure at the waterline is reduced to minimise primary wave forces so that a large submerged surface area is probably not required to counter-act the force due to radiation stress. Clearly, since the relative magnitudes of the various second order wave forces vary with the mean wave period it is only possible to play off one component against another, effectively, in certain sea states. But this may be possible in extreme sea states where mooring loads will almost certainly take their greatest values.

21. Typically, load deflection curves of compliant moorings for offshore structures exhibit an increasing stiffness with increasing extension, for example, catenary moorings. Therefore, a large movement of the structure due to a large wave on top of a slow drift back due to a low-frequency response will result in a high mooring load. It might, therefore, be expected that mooring loads would be reduced if moorings exhibited a decrease in stiffness with an increase in extension. The tube pump mooring, at present being studied as part of the wave power programme has such a characteristic. When this characteristic was model tested at HRS (19), it was found to produce a significant reduction in the mooring loads experienced with more conventional characteristics.

For example, a wave energy device 230m long on a multi-point mooring in a multi-directional sea with a significant height of 15m gave a maximum load of 705 tonnes force in one of the moorings when conventional characteristics were represented. A maximum load of 440 tonnes was obtained in the same mooring geometry with an optimum tube pump characteristic. Maximum movements of the centre of the device measured relative to the equilibrium position were about 38m for both characteristics. A spectral analysis of the mooring force record for the two characteristics (see figure 8) established that most of the reduction occurred

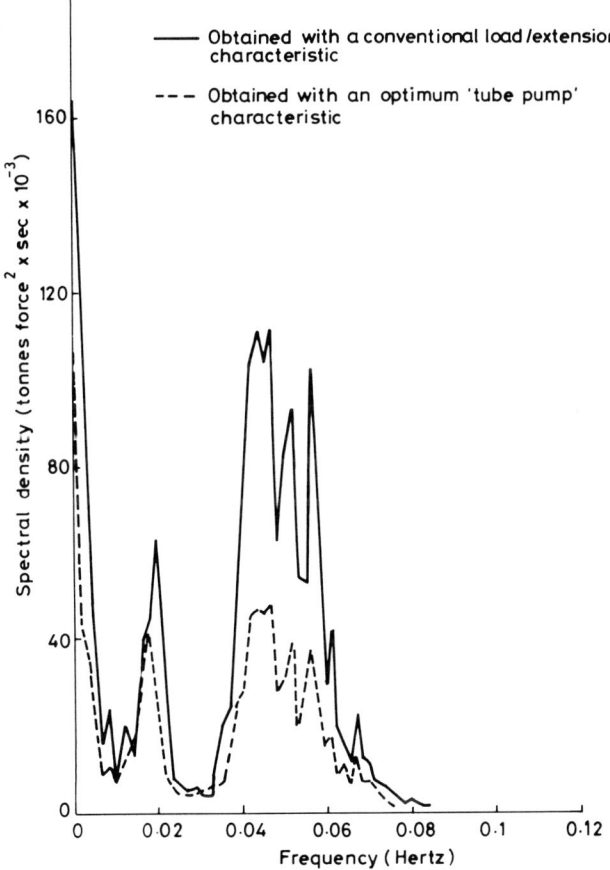

Fig 8.   Mooring line tension spectra in a 15m short-crested sea.

in the wave period component of the force. But figure 8 does show that some reduction was also obtained in the low-frequency component. Incidentally, these results were obtained with a tube pump characteristic that did not dissipate energy via hysteresis in the load/extension curve. When tests were carried out with energy dissipation in the moorings it was found that no additional reduction occurred in mooring loads.

WIND AND CURRENT LOADING

22. Wind and current forces are usually obtained from steady-flow experiments in a wind tunnel and by towing through calm water, respectively. The designer wishes to keep the number of model tests to a minimum, however, and so simple estimation procedures have been proposed. Isherwood(20), for example, analysed a large number of wind-tunnel measurements on merchant ships, and developed a set of coefficients for estimating wind forces on a wide range of vessels. Wind and current forces on offshore structures are often estimated by summing the drag forces on individual members, and making some allowance for interference effects. The accuracy of these procedures is not clear.

23. Vortex-shedding from structural members and mooring lines will cause oscillatory forces, both in-line and transverse to the current direction. The frequency of vortex-shedding f_v from a fixed cylinder of diameter D in a uniform flow velocity U is such as to make the Strouhal number approximately constant: $f_v D/U \approx 0.2$. If the vortex-shedding and natural response frequencies of the structure are close enough then there can be a large dynamic response (21). Fatigue damage may result.

24. Wind and current loads are often estimated independently of wave loading, and then superimposed. Waves and currents can interact, however, in a number of ways. Waves change their height, length and direction of travel when they encounter a current. A following current will cause the wavelength to increase and waveheight to fall; and vice-versa in an opposing current. In the latter case waves may exceed their limiting steepness and break (22). A current may also affect the way in which waves are diffracted by large structures. Drag forces on tubular members may also be affected: the force depends on the square of the combined (wave and current) velocity, and this is different from the sum of drag forces in separate wave and current conditions (16).

TWO EXAMPLES
Single-point moorings

25. The ship at a single-point mooring and the tethered
buoyant platform typify two different classes of moored
offshore structure. Concentrating first on the single-
point mooring, many of the design problems are associated
with the vessel's low-frequency response. Under certain
wind, wave and current conditions, pendulum-type sway motions
occur, with a period of perhaps 10-30 minutes, and coupled
with surge and yaw. These motions, often known as 'fishtailing' can cause severe loads in hawsers and connections.
These motions may arise in several different ways: through
an inherent instability of the system, non-linearities in
the hawser characteristic, variations in the wave drift
force, and, in some cases, wind gust loading.

26. Stability analysis. Linear perturbation methods can
help to delineate ranges of instability, and draw attention
to particular critical conditions. The system is considered
stable or unstable depending on whether the eigensolutions
represent a decaying or growing response. Faltinsen (23) and
Sørheim (24) thus identified several factors which affect
stability, including hawser length, damping, vessel draft
and trim, and propeller thrust. Astern thrust not only
stabilises the system, but also increases the mean hawser
load, making snatch loading less likely. The use of transverse thrusters, to change the vessel's heading, may sometimes improve stability. Also, the usual fishtailing
motions appear to be absent when rigid 'A'-frame yokes are
used. In model tests (25) of a large tanker on such a
single-point mooring in long crested random waves the ship
aligned itself with the wave direction and moved in pitch,
heave and surge only.

27. Numerical simulation. A linear stability analysis gives
no information about magnitudes of vessel response and hawser tensions. These may be estimated by numerical simulation
incorporating non-linearities in the hawser characteristic
and fluid loading. Many simulation programs exist, including
one currently being developed at NMI. Few have been validated very fully, and empirical 'corrections' are often
needed in order to obtain good agreement with experiment
(eg 26). Numerical difficulties often arise when the hawser tautens suddenly. Peak loads can be very sensitive to
small changes in the model, and in the way low-frequency
motions are combined with wave-frequency response of both
vessel and mooring tower. These components are often calculated separately. At times of peak loading the hawser
force can be highly non-linear, and simple linear super-

position of the separate motions can prove unreliable.

28. Modelling problems. Careful thought needs to be given
to the way in which environmental conditions are simulated
during either numerical or physical model testing. Wind
and current forces need to vary realistically with ship
heading angle. This often precludes use of simple weight-
and-pulley systems. Fans, mounted on the vessel and under
microprocessor control (27), offer one possible solution.
Fixed banks of fans placed around the vessel are also used
to represent wind loads. Such a system can be made to give
the correct wind speed profile above the water. The random-
ness, sequence length and stationarity of the wave history
are also important factors. NMI and HRS believe that at the
present time the only reliable means of obtaining design
data for ships at single-point moorings is by physical model
testing in very long sequences of random or psuedo-random
waves (see paragraph 19).

29. There is also evidence (25) that colinear wind, current and undirectional waves do not necessarily result in
the 'worst' design conditions. A transverse wave component
may cause roll, sway and yaw motions which significantly
affect peak hawser loads.

Tethered buoyant platforms (TBPs)

30. Forces and response motions over a wide range of frequencies affect TBP design. Typical ranges for a platform
in 200m water depth are shown in figure 9. Table 2 shows

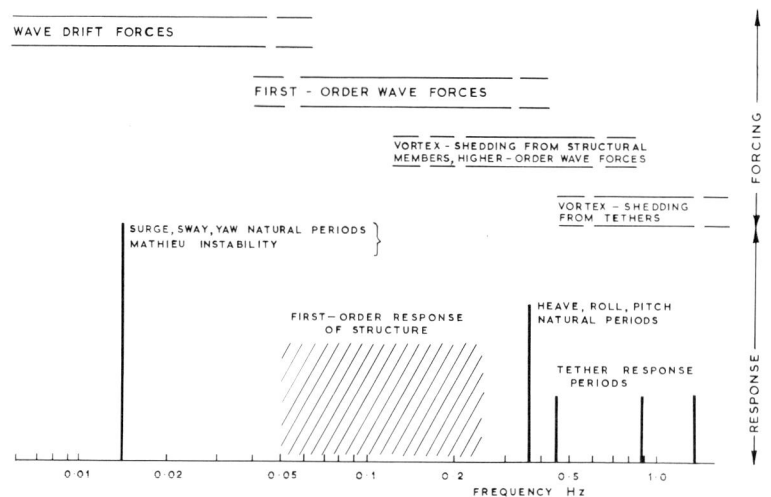

Fig 9. Typical forcing and response ranges for a TBP in 200m
water depth.

how the natural response periods vary with water depth (see also ref 28 for further similar calculations and discussion). The following values were assumed in making these estimates: $mg/T_o = 6$, $T_o/t_uA = 0.2$, yield stress $t_u = 1500MN/m2$, Young's modulus $E = 200GN/m2$, density $\rho = 9$ tonne/m^3, where T_o is the mean total tether tension, A is their total section area, ρA is their (mass + added mass) density per unit length. Vortex-shedding frequencies were estimated assuming a Strouhal number of 0.2, typical tether diameters 0.15 - 0.3m and a current speed 1m/s.

31. <u>Mean offset and low-frequency response</u>. The platform will surge, sway and yaw about a mean offset position, which will depend on the resultant of mean wind, current and wave drift forces. Natural surge, sway and yaw periods will generally lie well outside the range of wave activity, except perhaps in very exposed locations. Low-frequency motions will also be caused by slow variations in the wave drift force, perhaps by wind gusting or by a Mathieu form of instability.

32. Large Mathieu-type subharmonic motions have been recorded in ideal regular-wave conditions (29). There is doubt, however, about their occurrence and importance in irregular waves. Criteria proposed by Roberts (30) for an analogous problem in ship roll response suggest that, if the damping is linear, there is a critical value of damping below which unstable response can occur. The motion is always stable, in the sense that it remains bounded though perhaps large, if the damping is non-linear. It is usually difficult to esti-

Table 2. Typical natural periods for a tethered buoyant platform in water of various depths.

Response natural period	Formula	Typical periods (sec) when tether length ℓ =		
		100m	200m	400m
Surge T_s	$2\pi\sqrt{m\ell/T_o}$	49	69	98
Heave T_H	$2\pi\sqrt{m\ell/EA}$	1.9	2.7	3.8
Tethers, mode n T_n	$\dfrac{2\ell}{n}\sqrt{\rho A/T_o}$	$T_1 = 1.1$ $T_2 = 0.6$	2.2 1.1	4.4 2.2
Vortex-shedding T_v	$D/0.2u$	0.7 - 1.5		

mate the amount of damping present, and there are scaling problems in extrapolating from model-test results.

33. <u>Wave-frequency components</u>. First-order wave forces, response motions and tether tensions can be predicted well using either Morison's equation or wave diffraction theory, depending on the size and spacing of members. Figure 2, for example, compares amplitudes of the surge response and tether loads, calculated using the NMIWAVE program, with experimental values, over a range of regular wave numbers $k(=2\pi/\lambda)$ and wave heights.

34. <u>High-frequency response</u>. There are many possible sources of high-frequency forcing: for example very short and higher-order waves, wave breaking, non-linear forces associated with the platform's response and vortex-shedding. The magnitudes of these forces will usually be small, but they may cause resonant heave, roll, or pitch response and fatigue problems. In relatively shallow water, the designer will usually make the tethers stiff enough to keep the natural frequencies outside the range of wave activity. This becomes more difficult as the water depth increases. In order to maintain a constant natural period, both the total section area and length of the tethers have to increase, and there is a rapid escalation in both tether weight and cost. The designer may instead opt to allow a modest rise in the natural period, and rely on damping to avoid dynamic response and fatigue problems. Results in table 2 were obtained assuming the latter option.

35. <u>Tether response</u>. The tethers themselves may respond to vortex-shedding or other high-frequency forces. Only the lowest vibration modes are likely to be of any importance. Table 2 shows the natural periods of the first two modes, either of which may be excited by vortex effects.

CONCLUSIONS

36. Mathematical and physical models have complementary roles to play in design. This is particularly true of the design of moored systems, where the motions are often complex and non-linear, involve interactions between many different variables, and are often sensitive to small variations in any one of them. Perhaps the best use can be made of mathematical models in the early stages of design. For example, computational models like those developed by Standing (7) and Pinkster (8) (see paragraphs 12-15) will provide good estimates of wave loads and motions that occur at the wave period, and with recent developments these models now provide estimates of the low-frequency response as well as steady drift force effects. Such models help to give insight into the significance of various parameters, and enable the designer to vary conditions in a highly controlled and pre-

cise manner. Mathematical models can also help to provide
design data when physical scaling problems prove insuperable.
37. The mathematical models described above use the spectral approach and they offer a far more complete picture of
the response of moored structures than the deterministic
'design wave' approach. Further work is necessary, however, to calculate the effects of non-linear moorings and
non-linear damping. Therefore, after a moored structure has
been investigated using mathematical models it often makes
sense to model test the design in a wave tank. In some
cases testing in uni-directional random waves will provide
conservative estimates of structure response and such
results have a safety factor built into them, but there are
situations where multi-directional waves may provide more
critical conditions (25). The importance of directional
spread in the wave spectrum has been clearly demonstrated by
a comparison of orbital wave velocities measured in a storm
with predictions from a number of wave theories, both linear
and non-linear, where it was found that the linear multi-directional spectral model gave the best agreement. Therefore,
tests in multi-directional waves (see eg figure 10) can be

Fig 10. Tanker on rigid 'A'-frame yoke under test in a multi-
directional sea in the new offshore sea basin at HRS.

expected to provide a more realistic description of structure response. This should generate increased confidence in the results and lead, in the long term, to reduced safety factors that can be applied in the final design stage. However, to obtain accurate results the model basin must provide a good representation of the multi-directional wave spectrum, and the method of wave generation must be capable of providing a cycle of random waves that lasts a sufficient number of resonant periods to give a stable low-frequency response.

ACKNOWLEDGEMENTS

Part of this work was carried out for the Department of Energy and their support is gratefully acknowledged. The paper is published with the permission of the Directors of the Hydraulics Research Station and the National Maritime Institute.

REFERENCES
1.   FORRISTALL G.Z. et. al. Storm wave kinematics. Offshore Tech. Conf. paper 3227, 1978.
2.   SARPKAYA T. The hydrodynamic resistance of roughened cylinders in harmonic flow. Trans. Roy. Instit. Nav. Archit., 1978, 120, 41-58.
3.   STANDING R.G. Wave loading on offshore structures: a review. Nat. Mar. Inst. rep no. R102, 1981.
4.   PEARCEY H.H. The effects of surface roughness on the wave loading for cylindrical members of circular cross section. Nat. Mar. Inst. rep. no. R65, 1979.
5.   HOOFT J.P. A mathematical method of determining hydrodynamically induced forces on a semisubmersible. Trans. Soc Nav. Archit. Mar. Engrs., 1971, 79, 28-70.
6.   BOWERS E.C. Model simulation of ship movements. Dock and Harbour Authority, 1977, 58, 162-163.
7.   STANDING R.G., DACUNHA N.M.C. and MATTEN R.B. Slowly-varying second-order wave forces: theory and experiment. Nat. Mar. Inst. rep. to be published.
8.   PINKSTER J.A. Low-frequency second-order wave exciting forces on floating structures. NSMB rep. no. 650, 1980.
9.   LONGUET-HIGGINS M.S. and STEWART R.W. Radiation stresses in water waves: a physical discussion with applications. Deep Sea Res., 1964, 11, 529.
10. HYDRAULICS RESEARCH STATION. Second-order wave forces on wave power devices. Rep. EX 958, 1980.

11. BOWERS E.C. Long period oscillations of moored ships subject to short wave seas. Trans. Roy. Inst. Nav. Archit., 1976, 118, 181-191.
12. HANSEN N.E. et. al. Correct reproduction of long waves in physical models. 17th Int. Conf. Coastal Eng., Sydney, 1980.
13. BOWERS E.C. Long period disturbances due to wave groups. 17th Int. Conf. Coastal Eng. Sydney, 1980.
14. NEWMAN J.N. Second-order slowly-varying forces on vessels in irregular seas. Proc. Symp. Dyn. Mar. Vehicles Struct. in Waves. 1974, 182-186.
15. STANDING R.G., DACUNHA N.M.C. and MATTEN R.B. Mean Wave drift forces: theory and experiment. Nat. Mar. Inst. rep. no. R124, 1981.
16. PIJFERS J.G.L. and BRINK A.A. Calculated drift force on two semisubmersible platform types in regular and irregular waves. Offshore Techn. Conf. paper no. OTC2977, 1977.

17. LEAN G.H. Subharmonic motions of moored ships subjected to wave action. Trans. Roy. Instit. Nav. Archit., 1971, 113, 387.
18. DACUNHA N.M.C., HOGBEN N. and STANDING R.G. Responses to Slowly Varying Drift Forces and their Sensitivity to Methods of Wave Spectral Modelling: A Preliminary Assessment. Nat. Mar. Int. rep. R101, 1981.
19. HYDRAULICS RESEARCH STATION. Mooring forces on wave energy devices: effect of different load/extension characteristics. Rep. EX 1009, 1981.
20. ISHERWOOD K.M. Wind resistance of merchant ships. Trans. Roy. Instit. Nav. Archit., 1973, 115, 327
21. VERLEY R.L.P. and EVERY M.J. Wave induced vibrations of flexible cylinders. Offsh. Techn. Conf. paper no OTC 2899, 1977.
22. BURROWS R., HEDGES T.S. and MASON W.G. The influence of wave-current interaction on induced fluid loading. Proc. Symp. Hydrodyn. Ocean Engrg., Trondheim, 1981, 491-508.
23. FALTINSEN O.M., KJAERLAND O., LIAPNIS N. and WALDER-HAUG H. Hydrodynamic analysis of tankers at single-point mooring systems. Proc. 2nd Conf. Behaviour Offsh. Struct. (BOSS '79), London, 1979, 2, 177-206.
24. SØRHEIM H.R. Analysis of motion in single-point mooring systems. Norw. Mar. Res., 1981, 9, 2-13.
25. HUNTINGTON S.W. Wave loading in short-crested seas. Int. Conf. Waves and Wind Directionality with Application to Design of Structures, Paris, 1981.
26. WICHERS J.E.W. Slowly oscillating mooring forces in single-point mooring systems. Proc. 2nd Conf. Behaviour Offsh. Struct. (BOSS '79), London, 1979, 3, 661-692.

27. DAND I.W. Model studies of freely drifting and towed disabled tankers. RINA Symp. Behaviour Disabled Large Tankers, London, 1981.

28. MELLON B. and MILLER N.S. Tethered Buoyant Platforms: some design consideration and a review of some problem areas for further research. Glasgow Univ. rep. no. NAOE-HL-78-06, 1978.

29. ROWE S.J. and JACKSON G.E. An experimental investigation of Mathieu instabilities on tethered buoyant platform models. Nat. Mar. Inst. rep. no. R73, 1980.

30. ROBERTS J.B. The effect of parametric excitation on ship rolling motion in random waves. Nat. Mar. Inst. rep. no. R100, 1980.

2 A system for reduction of ship motions and hawser loads during offshore loading

O. ROPSTAD and H.-R. SØRHEIM, A/S Kongsberg Vapenfabrikk

Regularity and reliability of offshore loading operations are limited by motions resulting from large environmental forces. A single-point mooring system is analysed using mathematical models and simulation. Equilibrium, stability and natural frequencies are dealt with. On-line estimation of forces and motions is described, and a method for reduction of mooring loads and ship motions using the station-keeping capability of the ship is suggested.

1. INTRODUCTION

Due to the large cost of seabed pipelines, offshore loading has become an attractive alternative for transportation of oil and gas from offshore installations. There has been a strong development on the construction side, in order to meet the challanges of large water depths, hostile environmental conditions and increased demand for safety and reliability.

Far the most offshore loading systems are based on the single-point mooring principle. By this it is understood that the mooring consists of a single mooring line, called the hawser, between the ship bow and the loading terminal, or buoy.

Typical hawser lengths are 40-70 m. The crude is pumped through separate hoses.

The performance of the loading operation is strongly affected by the weather condition. During periods of strong wind and severe sea conditions the ship goes into a complex cyclic motion. This motion, in turn, leads to large tension in the hawser, and increased wear and tear of hoses and loading equipment. Under given conditions the loading must be abrupted, and the ship may have to unmoor and keep at a safe distance from the buoy until the weather conditions have improved.

An overall improvement in single-point mooring system performance should be expected if it is possible to reduce the motions during bad weather. Such a system is based on the assumption that tankers used for offshore loading are equipped for high manoeuvrability at near zero speed. By active control of the manoeuvring devices on the ship, both during the approach phase and loading operation, it should be possible to increase the overall safety and regularity of both operations and equipment.

In the following the static and dynamic behaviour of ships in a single-point mooring system is analyzed. The analysis is based on a mathematical description of the system. Finally a solution to the control problem is suggested, based on on-line state- and parameter estimation and modern control theory.

2. MATHEMATICAL MODELS

It is known from both full scale observations, [1], model scale experiments and theoretical analysis [2], that the relative motion between ship and buoy in single-point mooring systems consists of high and low frequency components. The high frequency motion is due to ship and buoy oscillating with wave frequencies, while the low frequency components contain the characteristic modes of the moored ship, originating from slowly varying environmental excitations, mooring system and maneuvering device operation.

The mathematical description is divided into models of the ship, the loading terminal, the mooring system and the environment. Coupling between ship and terminal is restricted to the mooring system, ship motion is assumed independent of water depth, and there is no interaction with other structures or ships.

2.1 The ship

High frequency model

First order ship motions in waves are assumed to be independent of the mooring system and low frequency ship motions, and described by standard hydrodynamic theory for ship motions in waves. From this theory the transfer function between wave height and ship motion is calculated for each relative wave direction and load condition (which in itself is a considerable numerical task). The actual ship response in irregular waves are calculated using the linearity assumption and superposition of responses in regular waves [3].

Low frequency model
a) General form of equations

The equation of motion in surge (longitudinal motion), sway (lateral motion), and yaw, (turning motion about the vertical axis) is described in a coordinate system (x,y,z) fixed in the ship's center section with positive x-axis forward. xy and xz are planes of symmetri and the ship is slender along the x-axis. Newton's law

29

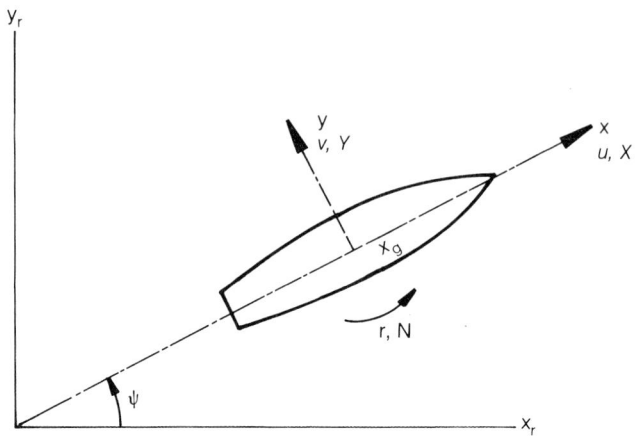

Fig. 1. Ship co-ordinates

of motion gives (a dot above a variable stands
for derivation with respect to time)

$$m(\dot{u} - vr - x_g r^2) = X$$

$$m(\dot{v} + ur + x_g \dot{r}) = Y \tag{1}$$

$$mk^2_{\psi\psi}\dot{r} + mx_g(\dot{v} + ur) = N$$

where (u,v,r) and (X,Y,N) are velocity and
force components in surge, sway and yaw respectively; m, mass, $k_{\psi\psi}$, radius of gyration in
yaw and x_g, position of center of gravity (see
also Figure 1). (u,v) is velocity relative to
the surrounding water. Let $(\dot{x}_r, \dot{y}_r)$ be the
ship's velocity in an earth fixed coordinate
system (x_r, y_r, z_r). Then

$$\dot{x}_r = u \cos \psi - v \sin \psi + v_{cu}\cos\psi_{cu}$$

$$\tag{2}$$

$$\dot{y}_r = u \sin \psi + v \cos \psi + v_{cu}\sin\psi_{cu}$$

where ψ is ship heading or rotation of the ship fixed coordinate system relative to the earth fixed system. (v_{cu}, ψ_{cu}) is velocity and direction in fixed coordinates of the ocean current.

The acceleration force is determined by the hydrodynamic reactions, maneuvering device operation, wind, waves and mooring hawser according to

$$X = X_{hydro} + X_{wind} + X_{wave} + X_{hawser}$$
$$+ X_{maneouvring}$$

$$Y = Y_{hydro} + Y_{wind} + \ldots\ldots$$

$$N = N_{hydro} + N_{wind} + \ldots\ldots$$

(3)

b) Hydrodynamic forces

Unlike ship motions in waves, where ideal flow theory is applicable with good results, low frequency motions of moored ships are significantly influenced by the viscous properties of the flow. Approximate theory and empirical relations, therefore, become necessary parts of these models. There exist two different approaches to this problem, a pure mathematical approach using Taylor series expansion, and a more rational physical approach based on a theoretical description of the flow mechanisms involved. Workable analytical expressions for different force components are derived under the assumptions of ideal flow and/or simplified hull geometry. Empirically determined correction factors can be included to account for actual observed ship behaviour. The present model is based on these principles.

The hydrodynamic force consists of the ideal flow force and forces due to the viscosity of the flow.

Ideal flow forces, [4]

$$X_{id} = -a_{xx}\dot{u} + a_{yy}vr + a_{y\psi}r^2$$

$$Y_{id} = -a_{yy}\dot{v} - a_{xx}ur - a_{y\psi}\dot{r} \qquad (4)$$

$$N_{id} = -a_{\psi\psi}\dot{r} - (a_{yy} - a_{xx})uv - a_{y\psi}(\dot{v} + ur)$$

where a_{ij} are the so-called added mass coefficient in the i-direction due to motion in the j-direction.

Calculations of added mass are usually based on two-dimensional methods, (slender ship approximation), and three-dimensional correction factors.

Viscous forces consist of hydrodynamic lift, cross-flow drag and frictional drag. For small angles of attack, i.e. $|v| \ll |u|$ and $|r|L \ll |u|$ the lateral viscous force is dominated by <u>viscosity generated lift</u>,[5]

$$Y_{lift} = -K_{\ell}\rho_w \pi T^2 (|u|v + x'_{p1}Lur)$$

$$\qquad (5)$$

$$N_{lift} = -K_{\ell}\rho_w \pi T^2 L(x'_{p1}uv + x'^2_{p2}L|u|r)$$

where ρ_w is density of water, L ship length, T mean draft and K_{ℓ}, x'_{p1}, x'_{p2} non-dimensional empirical coefficients describing the distribution of lift along the hull. The viscous model in surge is assumed to consist of <u>frictional drag</u> [6]

$$X_{visc} = -\tfrac{1}{2}\rho_w C_{dx}S|u|u \qquad (6)$$

where C_{dx} is frictional drag coefficient and S area of the under-water hull.

Similarly to lift, a two-dimensional <u>cross-flow drag</u> description gives, [7]

$$Y_{cross-flow} = \begin{cases} -\tfrac{1}{2}\rho_w C_{dy} TL \; sign \, (v) \, [v^2 + r^2 L^2/12] \\ \qquad\qquad\qquad\qquad |x_p| \geq \tfrac{1}{2}L \\ \rho_w C_{dy} T [x_p v^2 + (x_p^2 - \tfrac{1}{4} L^2) vr + \tfrac{1}{3} x_p^3 r^2] \\ \qquad sign(r) \qquad |x_p| < \tfrac{1}{2}L \end{cases}$$

$$N_{cross-flow} = \begin{cases} -1/12 \; \rho_w C_{dy} TL^3 |v|r \qquad |x_p| \geq \tfrac{1}{2}L \\ \tfrac{1}{2} \; {}_w C_{dy} T[(x_p^2 - \tfrac{1}{4} L^2) v^2 + \tfrac{4}{3} x_p^3 vr \\ + \tfrac{1}{2} (x_p^4 - 1/16 \; L^4) r^2] sign(r) \end{cases}$$

$$|x_p| < \tfrac{1}{2}L \quad (7)$$

x_p is the position of zero effective cross-flow, $x_p = -v/r$, and C_{dy} the mean lateral drag coefficient determined from a three-dimensional model test in lateral flow.

c) Maneuvering device forces

In addition to conventional maneouvring equipment, main propulsion and rudder, the ship is assumed to have fixed bow and stern lateral thrusters. Bow thruster is accepted as an indispensable tool on offshore loading tankers in order to increase maneuvering near the terminal. Lately, a tanker has been contracted which employs both bow and stern thruster.

33

All maneuvering device reaction forces are modelled by their respective steady state force characteristics based on model tank tests. Propulsion and rudder forces are coupled, while the lateral thrusters are assumed not to interact with each other, or with the other units. The effect of ship velocity is included, however, the effect is small for most actual motions. In particular, rudder efficiency is dominated by the main propeller action, which is frequently a fixed astern thrust, and is therefore practically useless as a maneouvring device in moored condition.

d) Wind forces

The wind model assumes the wind force to be a viscous drag force, and is based on the empirical drag coefficients $C^{wind}(\alpha)$ in surge, sway and yaw, [6]

$$X_{wind} = \tfrac{1}{2}\rho_a C_x^{wind}(\alpha) A_T V^2$$

$$Y_{wind} = \tfrac{1}{2}\rho_a C_y^{wind}(\alpha) A_L V^2 \tag{8}$$

$$N_{wind} = \tfrac{1}{2}\rho_a C_N^{wind}(\alpha) A_L L V^2$$

where ρ_a is density of air, A_T and A_L transverse and lateral exposed wind area, α, angle of attack and, V, relative velocity of wind, respectively.

If (v_{wi}, ψ_{wi}) is the wind velocity in fixed coordinates, then $\alpha = \psi_{wi} - \psi$ and $V^2 \approx v_{wi}^2$, neglecting the effect of ship's own velocity.

e) Wave forces

The mathematical description of slowly varying wave forces is questioned. For many applications the method suggested by Hsu and Blenkarn, [8], has been employed. They assume that the forces and moments in irregular waves can be calculated from the second order transfer function in regular waves, R, according to

$$F_i^{wa} = \tfrac{1}{2}\rho g R^2 (T_i) \zeta_i^2 L \tag{9}$$

where F_i^{wa} is the second order wave force, g the acceleration of gravity and T_i and ζ_i the instantaneous regular wave period and wave amplitude respectively.

R usually depends on the relative wave heading in a complex way. In this model, a simplified symmetrical relation is used

$$X_{wave} = F_x^{wa} |\cos(\psi_{wa}-\psi)| \cos(\psi_{wa}-\psi)$$

$$Y_{wave} = F_y^{wa} |\sin(\psi_{wa}-\psi)| \sin(\psi_{wa}-\psi) \tag{10}$$

$$N_{wave} = F_\psi^{wa} \sin 2(\psi-\psi_{wa})$$

where ψ_{wa} is wave direction and F_x^{wa}, F_y^{wa} and F_ψ^{wa} the surge, sway and yaw wave force in head, beam and bow waves respectively. It is emphasized that the errors involved using (10) may be quite large.

d) Mooring forces

The mooring system, consisting of a single mooring hawser between the ship bow and the buoy, is modelled by its static force-elongation characteristic, f_h, and length in unloaded condition, ℓ_{oo}. Hawser tension force is

$$f = \begin{cases} 0 & \ell \le \ell_{oo} \\ f_h\left(\dfrac{\ell-\ell_{oo}}{\ell_{oo}}\right) & \ell > \ell_{oo} \end{cases} \tag{11}$$

35

f_h is usually very nonlinear and is modelled by an interpolating curve.

The mooring force, F, acts at the bow in position: x = a, Figure 2

$$X_{hawser} = F \cos \gamma$$

$$Y_{hawser} = F \sin \gamma \qquad (12)$$

$$N_{hawser} = aF \sin \gamma$$

γ is the hawser direction relative to the ship heading.

2.2 The loading terminal

The variety of loading terminals is such that a general type model of fluid reaction forces on the terminal is not practical. The articulated loading platform (ALP) is used in this study. The ALP is a bottom pivoting, rigid and slender tower with a buoyancy module which gives vertical stability, Figure 3.

Assuming small buoy excursions compared to water depth and wavelength, and negligible wind drag, the equation of motion of the buoy in a surge at hawser level becomes

$$m_b \ddot{x}_b + k_b x_b = X_b^{hydro} - F \cos (\gamma + \psi) \qquad (13)$$

where m_b and k_b are the equivalent buoy mass and restoring coefficient respectively. The equation for sway is quite similar, and all other modes are suppressed. The hydrodynamic force term is assumed to consist of hydrodynamic added mass reaction force, first order wave excitation and viscous drag, and is calculated using strip theory and Morrisons equat-

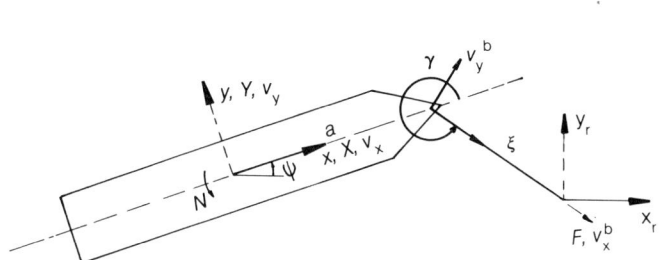

Fig. 2. Moored ship co-ordinates

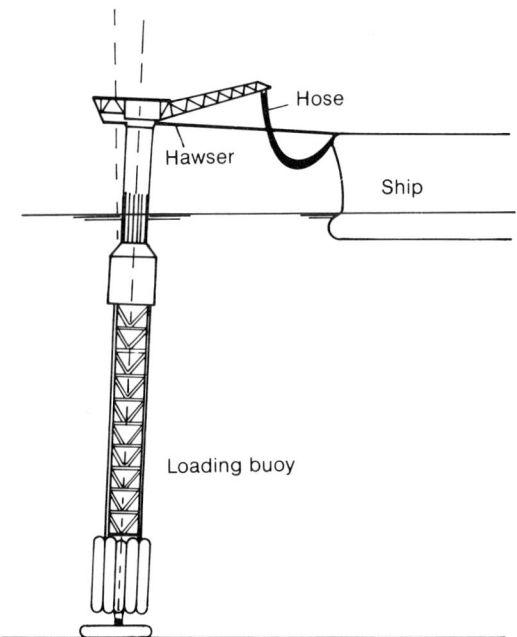

Fig. 3. Single-point mooring with ALP

ion at each strip. Let the force at level z be $X_{bz}(z)$. Then

$$X_{bz}(z) = \rho_w \frac{\pi D_b^2(z)}{4} (1+C_{Mb}(z))a_{wa}(z) \cos \psi_{wa}$$

$$+ \frac{1}{2}\rho_w C_{Db}(z) D_b(z) |v_{wa}(z)\cos\psi_{wa} + v_{cu}\cos\psi_{cu}$$

$$- \frac{z}{H}x_b| (v_{wa}(z) \cos\psi_{wa} + v_{cu} \cos\psi_{cu} - \frac{z}{H}x_b)$$

and

$$X_b^{hydro} = \frac{1}{H} \int_0^{HWL} X_{bz}(z) z \, dz \qquad (14)$$

where a_{wa} and v_{wa} are the horizontal wave particle acceleration and velocity respectively, D_b, C_{Mb} and C_{Db} are buoy diameter, added mass coefficient and drag coefficient respectively, H is hawser level and HWL water depth.

2.3 The environment

Wind is modelled as two stationary, zero-mean gaussian gust processes superimposed on a steady state mean wind. Each process has the form

$$\dot{q} = c_1 q + c_2 z \qquad (15)$$

$$v_g = c_3(q+z)$$

where v_g is the gust velocity, z a zero-mean gaussian, white noise process with specified variance and (c_1, c_2, c_3) model coefficients. These are calculated by fitting the model to a chosen turbulence spectrum [9].

Irregular waves are described by a sum of regular, small amplitude waves with random phase. Assuming deep water and first order wave theory. The wave elevation is

$$\zeta(t) = \sum_{i=1}^{m} \zeta_i \cos(\omega_i t + \varepsilon_i) \qquad (16)$$

where the amplitude ζ_i and frequency ω_i of each wave component is calculated from a typical ocean wave spectrum.

Current has a constant value and a specified vertical profile.

3. ANALYSIS OF MOTION IN SINGLE-POINT MOORING SYSTEMS

3.1 High frequency motion

As already stated the first order wave response of the ship is assumed equal to that of a free-floating ship. The oscillation of the buoy with the waves may, however, be significantly affected by the moored ship. For large tankers the equivalent buoy inertia,(ALP), is small compared to the ship displacement mass, and the buoy observes the ship as a slowly changing restoring force. Normally damping is very small and outside resonance the following transfer function yields a good description of the buoy excursion

$$\frac{x_b}{F_{ex}}(\omega) = \frac{1}{k_b + k_h} \cdot \frac{1}{1-(\omega/\omega_c)^2} \quad \omega \neq \omega_c \qquad (17)$$

where ω_c is the slowly changing natural frequency due to the hawser restoring coefficient k_h.

A typical area of responses to waves is shown in Figure 4.

3.2 Low frequency buoy motion

It is known that the motion of buoyant structures like the ALP may become unstable when the wave period is twice the resonance period, [10].

Such conditions would require extremely long
waves and large hawser tension, and is not likely to occur.   Normally the buoy behaves
like a simple restoring element following the
slowly changing ship bow motion.

3.3 Low frequency ship motion

Low frequency ship motions are due to low frequency excitations from the environment and
locally unstable mooring conditions.   Due to
the dimension and nonlinear character of the
model, any explicit form analytical results involve some sort of simplification and restrict-
ion on the models and the solutions.   A more
general analysis must be based on the study of
solution curves obtained from numerical   equations.   In the following the equilibrium conditions of mooring configurations and a simplified
set of equations of motion are derived.
The motion near equilibrium   is characterized
by the eigenvalues which under different assumptions leads to natural frequencies and stability criteria.

a) Equilibrium positions

When the ship is supported by a single mooring
force, the determination of equilibrium positions can be carried out sequentially starting
with the equilibrium ship heading.
Let index 0 denote values at equilibrium.   Ship
heading, ψ_0, is determined by the equation of
moment equilibrium at the mooring hawser connection point

$$N(\psi_0) - aY(\psi_0) = 0 \tag{18}$$

This is a nonlinear equation which usually has
to be solved numerically.   Only solutions

$\psi_0 \in [-\pi/2, \pi/2]$ are of interest, Figure 2. When ψ_0 is calculated, F_0 and Υ_0 are determined from $X(\psi_0)$, and $Y(\psi_0)$, and additional forces and positions by using the mooring force character-istics and trigonometric relations.

ψ_0 is determined by environmental forces and lateral maneuvering forces. Note that astern propulsion does not affect ψ_0, but can be used to reduce Υ_0. Similarly ψ_0 can be controlled by a lateral thruster, preferably located as near the stern as possible. A bow thruster, on the other hand, is practically useless for con-trolling mean ship heading.

b) Linearized analysis of motion

In order to simplify the equation of motion and obtain an explicit expression for the current forces, a new set of linear position and vel-ocity coordinates, (ξ, Υ) and $v_x^b, v_y^b)$ are intro-duced. (ξ, Υ) is the mooring line length and direction relative to the ship heading, and (v_x^b, v_y^b) is the longitudinal and lateral velocity of the mooring hawser bow connection point, Figure 2. Note that ξ includes both the hawser length and buoy excursion.

Assume, for simplicity, that $x_g = d_{\chi\psi} = 0$ in (1) and (4), and let $x_p^T = [\xi, \Upsilon, \psi]$ and $x_v^T = [v_x^b, v_y^b, r]$ denote pertubations around equilib-rium values. The linearized equation of motion then becomes [9]

$$
\begin{bmatrix} \dot{x}_p \\ \dot{x}_v \end{bmatrix} = \begin{bmatrix} 0 & A_{12} \\ A_{21} & A_{22} \end{bmatrix} \begin{bmatrix} x_p \\ x_v \end{bmatrix}
\tag{19}
$$

where A_{12}, A_{21} and A_{22} are matrices which dep-end on the equilibrium position $x_{po}^T = [\xi_o, \Upsilon_o, \psi_o]$.

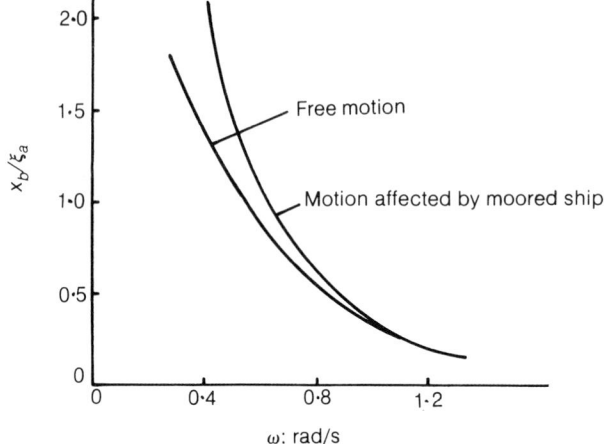

Fig. 4. ALP response curves in waves

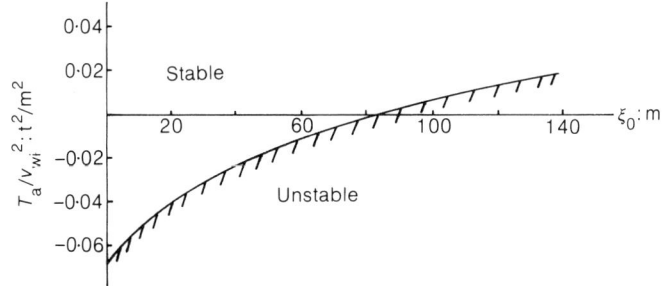

Fig. 5. Local stability condition in wind for ballasted ship

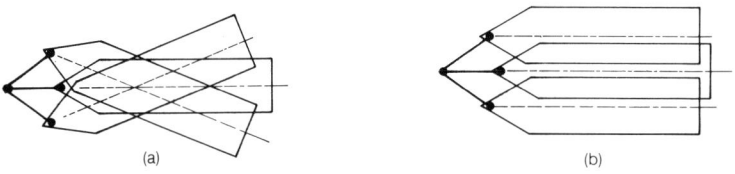

Fig. 6. Characteristic motion of moored ship: (a) fish tailing; (b) galloping

Here A_{12} describes the kinematics, A_{21} restoring and A_{22} damping properties, respectively. In order to obtain an analytical solution to the sixth order eigenvalue problem assosiated with (19), some further simplifications are made

i)   no damping
ii) longitudinal symmetry
iii) no environmental forces

The elements of A_{12} and A_{21} are given in Appendix. No damping, i), yields $A_{22} = 0$ while ii) or iii) yields particular simple forms of A_{21}.

c) Simplified stability theory

Destabilizing forces are due to environmental turning moments, and their derivatives with respect to ship heading. Because these forces are largest when the force is acting along the ship's longitudinal axis, one may expect the longitudinal symmetry to describe a worst case condition. Consider this case. Then, the surge mode, which is always stable, decouples and the sway-yaw part of the characteristic polynomical becomes

$$\lambda^4 + A\lambda^3 + B\lambda^2 + C\lambda + E = 0 \qquad (20)$$

Local stability is achieved for $A>0$, $B>0$, $C>0$, $E>0$ and $ABC - C^2 - A^2E>0$, which needs numerical evaluation. It is shown, however, [9], that for the undamped case the critical condition is expressed by, $(B>0)$

$$B^2 - 4E > 0$$

which, upon working out the expressions, yields a relation between ξ_0 and F_0 that must be satisfied for a locally stable system.

Stability for any mooring line length ξ_0 is obtained when

$$F_0 > F_0^{max} = \frac{I_z/M_y \; Y_\psi^{en} + aN_\psi^{en}}{I_z/M_y + a^2} \qquad (21)$$

or, if $F_0 < F_0^{max}$, when

$$F_0 > F_0^{crit} =$$

$$\xi_0 \left\{ \frac{[I_z/M_y(aY_\psi^{en}-N_\psi^{en})]^{\frac{1}{2}}+[a[I_2/M_y Y_\psi^{en}+(\xi_0+a)N_\psi^{en}]]^{\frac{1}{2}}}{I_z)M_y+a(\xi_0+a)} \right\}^2$$

$$(22)$$

where M_y and I_z are the virtual masses and (Y_ψ^{en}, N_ψ^{en}) external force derivates in sway and yaw respectively. Note that in the symmetrical case the hawser tension is the algebraic sum of longi- tudinal environmental force, X^{en}, and astern thrust, T_a, i.e. $F_0 = X^{en} + T_a$, and (21) and (22) is a relation between astern thrust and mooring li- ne length. The stability requirement (22) is freq- uently too strong, and the result for a typical 100 000 dwt tanker in ballast condition and wind exitation is shown in Figure 5.

When the ship becomes locally unstable, it goes into stable oscillations in limit cycles, with a typical fishtailing and galloping motion behav- iour, Figure 6. The amplitude is determined by the nonlinear properties of the exitation, damp- ing and mooring system forces.

d) Resonance properties

In most cases, except some involving strong cur- rent, the eigenvalues are complex conjugated in pairs, defining three natural frequencies of which at least two have small relative damping. The motion response is therefore dominated by these frequencies.

For the symmetrical case with locally stable eigenvalues and E<<B, in (20), the resonance

frequencies are [9]

$$\omega_1 = \sqrt{q_x k}$$

$$\omega_2 \approx \sqrt{(\frac{q_y + a^2 q_\psi}{\xi_0} + aq_\psi)F_0 - q_\psi N_\psi^{en}}$$ (23)

$$\omega_3 \approx 0$$

where k is the low frequency mooring system el-asticity and $q_x = I/M_x$ etc.

If the environmental forces are acting from different angles, or a nonsymmetric maneouvring load is applied, the longitudinal symmetry is lost. An approximate solution to the eigenvalue problem is now obtained under the assumption that the excitation force derivatives w.r.t. ship heading are small. Thus, neglection damping and excitation force derivatives, it can be shown that

$$\omega_1 \approx \sqrt{q_{11}(\gamma_0)k}$$

$$\omega_2 \approx \sqrt{(\frac{q_{22}(\gamma_0)}{\xi_0} + q_{23}(\gamma_0))F_0}$$ (24)

$$\omega_3 = 0$$

where $q_{ij}(\gamma_0)$ are defined in the appendix. The expression (24) is exact for $\gamma_0 = 0$, which indi-cate the effect of environmental force deriva-tives upon ω_2^2 to be $-q_\psi N_\psi^{en}$. Numerical calcu-lations show that ω_1 is well predicted while ω_2 and ω_3 are over- and underestimated respectively. The variation of ω_i^2, $i = 1,2,3$, with γ according to (24) is shown in Figure 7.

In figure 8 a typical hawser tension response to 20 m/s wind 5.5 m wave and 0.25 m/s current, waves acting $45°$ to wind and current, is exam-plified.

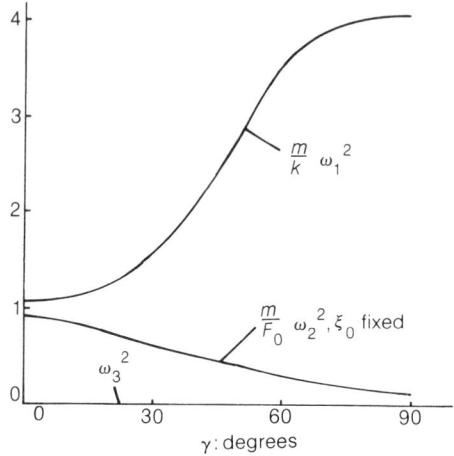

Fig. 7. Variation of approximate natural frequencies of oscillation with hawser angle

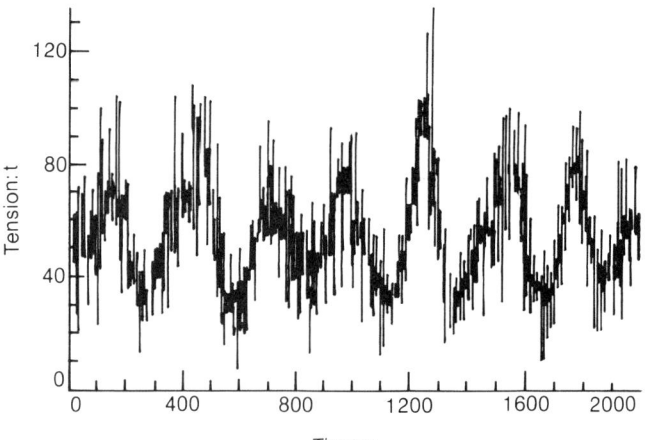

Fig. 8. Simulated hawser tension

4. ESTIMATION

In order to design an efficient control system for the ship, feed-back from the complete low frequency motion state is necessary. This information can be provided using different techniques. A modified Kalman filter is used here.

4.1 Measurements

Input to the filter consists of six measurements $y_1 \ldots y_6$ where

(y_1, y_2): relative position between ship and buoy

y_3: ship heading

y_4: hawser tension

(y_5, y_6): wind velocity and direction

The position measurements, (y_1, y_2, y_3), are essential in order to achieve observability of a motion model. Hawser tension, y_4, represents an important information by itself, and yields better dynamic filter properties. Wind measurement, (y_5, y_6) allows for an approximate instantaneous compensation of the wind force.

4.2 Filter models

a) Wave noise suppression

A well known problem in dynamic positioning of vessels is the suppression of wave induced high frequency components in the low frequency estimates. This problem is solved by locking a simple harmonic high frequency oscillator to each measurement. The oscillator model is

$$\dot{y}_{1Hi} = y_{2Hi}$$
$$\dot{y}_{2Hi} = -\omega_{Hi}^2 y_{1Hi} \qquad i = 1, \ldots, 4$$

(25)

where y_{1Hi} is the rapidly varying measurement
components due to buoy and ship motion with
wave frequencies and ω_{Hi} is the predominant frequency of oscillation. By adding this model in
the measurement prediction, the high frequency
components are effectively suppressed in the
innovation signal. This method was originally
suggested by Balchen et al. [11].

b) Low frequency motion

An important aspect of the present filter model
for slowly varying motion has been to take into
account the particular mooring system of a single-point mooring system. The following state
vector has been adopted

$$x_L^T = [\xi_E, \Upsilon_E, \psi_E, v_{xE}, v_{yE}, r_E]$$

where $(\zeta_E, \Upsilon_E, \psi_E)$ are the filter equivalent of
the position coordinates used in section 3,
and (v_{xE}, v_{yE}, r_E) are the velocity over ground
in ship fixed coordinates of the ship's center
of gravity.

The mooring system is modelled as a linear
spring containing the elastic properties of
both the hawser and the buoy.

Since wave and current are unknown, only their
mean equivalent effect in each degree of freedom is observable from the available measurements. Wind forces, however, are taken into
account in terms of a mean and a slowly varying part. A reasonable way to do this is shown
in [12]. The model is

Kinematics

$$\dot{\xi}_E = -v_{xE} \cos\Upsilon_E - (v_{yE} + ar_E)\sin\Upsilon_E$$

$$\xi_E \dot{\gamma}_E = v_{xE} \sin\gamma_E - (v_{yE} + ar_E) \cos\gamma_E - \xi_E r_E \quad (26)$$

$$\dot{\psi}_E = r_E$$

Dynamics

$$M_x \dot{v}_{xE} = a_1 v_{yE} r_E + a_2 v_{xE} + X_E^{wind} + F_E \cos\gamma_E$$

$$+ X_E^{bias} + X_E^{man}(u)$$

$$M_y \dot{v}_{yE} = b_1 v_{xE} r_E + b_2 v_{yE} + Y_E^{wind} + F_E \sin\gamma_E$$

$$+ Y_E^{bias} + \gamma_E^{man}(u)$$

$$I_z \dot{r}_E = c_1 v_{xE} v_{yE} + c_2 r_E + N_E^{wind} + aF_E \sin\gamma_E$$

$$+ N_E^{bias} + N_E^{man}(u) \quad (27)$$

Mooring force

$$F_E = k_E(\zeta_E - \zeta_{oE}) + F_{oE} \quad (28)$$

In (27) the coefficients a_i, b_i and c_i, $i=1,2$, are hydrodynamic ship parameters which are fixed for a given load condition. X_E^{wind} is the wind force estimate in surge using the wind-drag coefficient in the form shown in (8), X_E^{bias} includes the mean wind and current force and any force deviations due to incomplete modelling, $X_E^{man}(u)$ the controllable maneuvering force, etc.

4.3 Estimator design

Using a modified extended Kalman filter (EKF) on the measurements $y_1 .. y_4$ the following 18 states are estimated recursively

$\hat{X}_{H1} - \hat{X}_{H8}$: rapidly varying oscillator states,
$$y_{1Hi}, \ y_{2Hi}, \ i = 1,..4$$

$\hat{X}_{L1} - \hat{X}_{L6}$: slowly varying tanker motion, ξ_E etc.

$\hat{X}_{L7} - \hat{X}_{L9}$: bias force in surge, sway and yaw, X_E^{bias} etc.

$\hat{X}_{L10}$: mean hawser tension, F_{0E}

Using a tracking filter based on innovation sensitivities, [13], the following five slowly varying parameters are estimated recursively

$\hat{\theta}_1 - \hat{\theta}_4$: wave filter frequencies, ω_{Hi}, i=1,..4

$\hat{\theta}_5$: mooring system elasticity, k_E

The extended Kalman filter is given by Jazwinski [14]. The nonlinear form of the filter model yields a time-varying feed-back gain K. However, the heavy computational load associated with the computation of K is not acceptable in practical applications, which has lead to the development of a modified filter. In short, the modification consists of using the stationary values of K assuming nominal parameters and environmental excitations. This is a highly practical and good soultion when the sensivity to changing operating conditions and parameters are small. In this particular case, however, this was not so for the equilibrium hawser angle, Y_{0E}, and the feed back is therefore a function of this parameter.

The parameter tracking filter has the following form

$$\hat{\theta}_{k+1} = \hat{\theta}_k - \alpha(\frac{\partial\varepsilon_k}{\partial\hat{\theta}})^T R^{-1}\varepsilon_k \tag{29}$$

where α is an adjustable factor, ε_k innovation, R innovation covariance and $\partial\varepsilon_k/\partial\hat{\theta}$ the sensivity of the innovaton to the parameter. The calculation of $\partial\varepsilon_k/\partial\hat{\theta}$ may be a considerable task. For this system, however, there is one independent model associated with each parameter, and, most conveniently, the solution for the wave filter frequency sensitivities have been found analytically on explicit form, [12]. Using a similar technique, an analytic solution to $\partial\varepsilon_k/\partial\hat{\theta}_5$ is obtained assuming the major information with respect to $\hat{\theta}_5$ is contained in the innovations ε_1 and ε_4.

The wind force is handled independently using the wind measurements y_5 and y_6.

The filter structure is shown in Figure 9 and a typical set of estimates in Figure 10.

5. CONTROL

The main control objective is to reduce the relative low frequency motion between the loading buoy and the tanker bow. Typically the tanker is allowed to settle in any equilibrium position. This correspond to a so-called "minimum thrust" configuration, i.e. $E(u) = 0$ for all maneuvering units, and the hawser takes up the mean environmental load. If a preferred position is wanted, a fixed reference position is specified.

The feed-back has the following general form

$$u = u_{PD} + u_I + u_{FF} \tag{30}$$

where the notation PD = proportional + derivative, I = integral and FF = feed-forward is used. Feed-forward covers the possibility to react

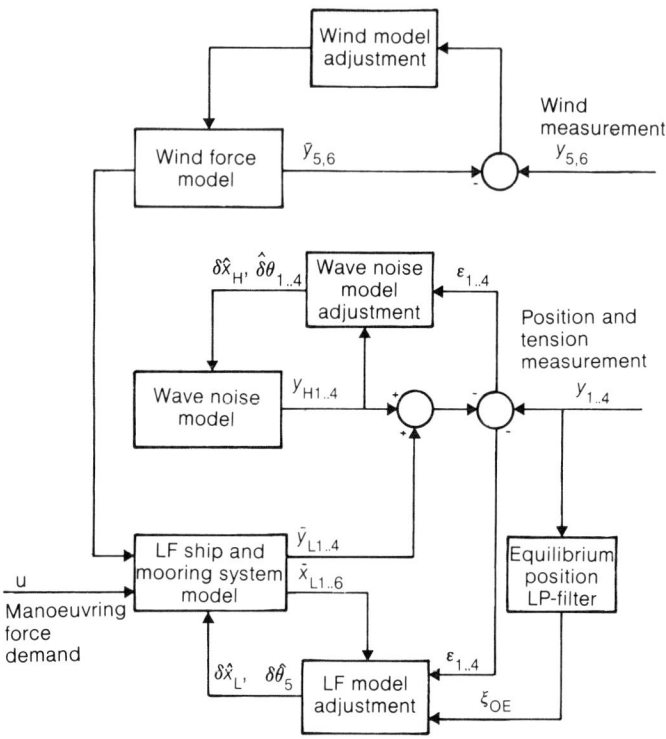

Fig. 9. Estimator structure

instantaneously to sudden changes in the wind force. u_{PD} is designed using a quadratic performance criterion and optimal control theory with the additional contraint that the relative damping is around 0.7 under all operating conditions. Due to the strong state nonlinearities, the result is made a function of the mooring configuration, using the method described for the filter gain. Thus

$$u_{PD} = G(\gamma)(\hat{x}_L - x_{ref}) \tag{31}$$

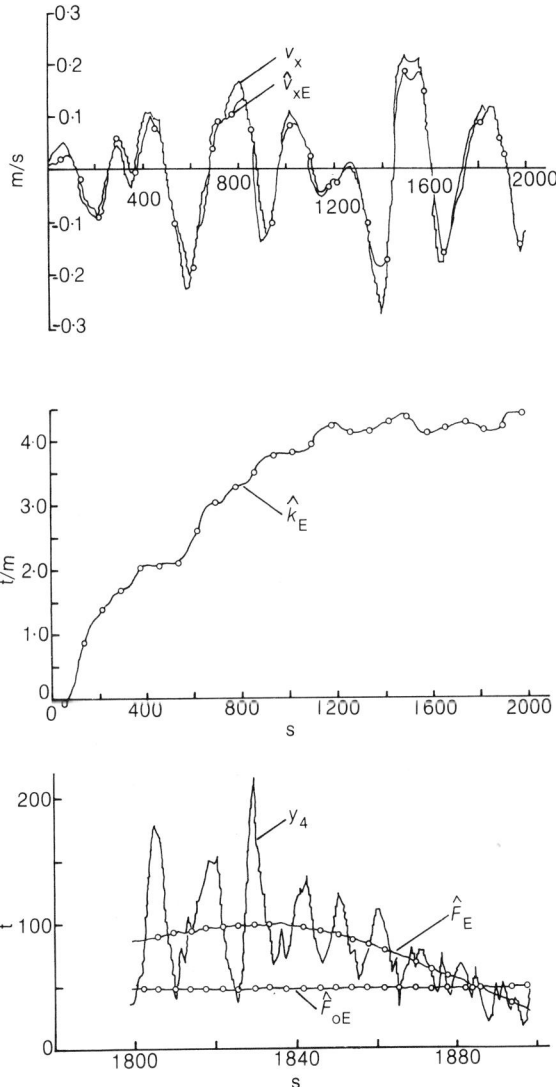

Fig. 10. Estimator states: (a) surge velocity over ground;
(b) mooring system elasticity; (c) hawser tension

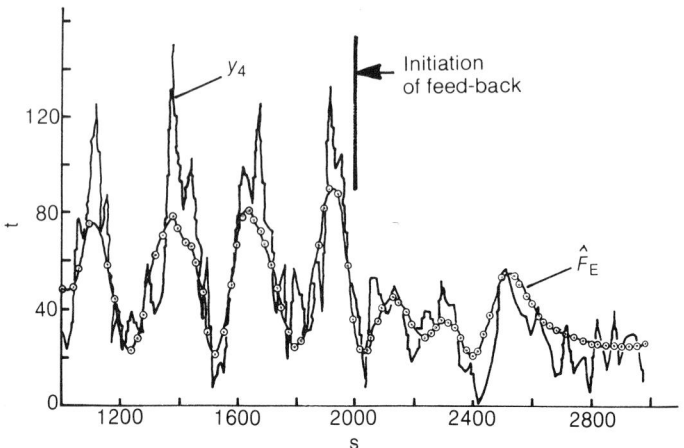

Fig. 11. Effect of feedback on hawser tension

where x_{ref} is either a specified value, or the equilibrium condition evaluated by low-pass filtering of the position measurements. u_I and u_{FF} are unit feed-back of the mean accelerating force estimate and the wind force estimate respectively [12]. The reduction in hawser tension amplitudes is indicated in Figure 11. Feedback is used in the last half section of the record. It is assumed above that all three degrees of freedom of the ship, surge, sway and yaw, are controllable, e.g. from the main propeller and lateral thrusters at bow and stern. If the stern thruster is missing, it should still be possible to control bow motion surge and sway, however, the ship heading would no longer be controllable.

6. CONCLUSION

The bow moored tanker in a single point mooring system experiences slowly varying motions due to forces from wind, waves and current. Equilibrium positions may be locally unstable. It is possible to reduce these motions and thereby reduce the excitations, wear and tear of critical equipment like mooring hawser and loading hoses by automatic control of the ship motions. This requires, at least, a bow thruster and the main propeller.

Appendix

$$A_{12} = \begin{bmatrix} -1 & 0 & 0 \\ 0 & -\dfrac{1}{\xi_0} & -1 \\ 0 & 0 & 1 \end{bmatrix}$$

$$A_{21} = \begin{bmatrix} q_{11}k, & q_{12}F_0, & q_x X_\psi^{en} \cos \gamma_0 + (q_y Y_\psi^{en} + aq_\psi N_\psi^{en}) \sin \gamma_0 \\ \\ q_{21}k, & q_{22}F_0, & -q_x X_\psi^{en} \sin \gamma_0 + (q_y Y_\psi^{en} + aq_\psi N_\psi^{en}) \cos \gamma_0 \\ \\ q_{31}k, & q_{32}F_0, & q_\psi N_\psi^{en} \end{bmatrix}$$

$$A_{22} = \{a_{22_{ij}}\} \qquad F_\psi^{en} = [X_\psi^{en}, Y_\psi^{en}, N_\psi^{en}]^T$$

$Q = \{q_{ij}\}$ is symmetric with

$$q_{11} = q_x \cos^2 \gamma + (q_y + a^2 q_\psi) \sin^2 \gamma$$

$$q_{12} = \tfrac{1}{2}(q_y + a^2 q_\psi - q_x) \sin 2\gamma$$

$$q_{13} = aq_\psi \sin \gamma$$

$$q_{22} = q_x \sin^2 \gamma + (q_y + a^2 q_\psi) \cos^2 \gamma$$

$$q_{23} = a q_\psi \cos \gamma$$

$$q_{33} = q_\psi$$

$$q_x = 1/M_x$$

$$q_y = 1/M_y$$

$$q_\psi = 1/I_z$$

REFERENCES

1. HARING R.E. Single-point tanker mooring measurements in the North Sea. Proc. Offshore Technol. Conf., 1976, Houston, 2711.
2. PINKSTER J.A. and REMERY G.F.M. The role of model tests in the design of single point mooring system. Proc. Offshore Technol. Conf., 1975, Houston, 2212.
3. GALTUNG F.L. and LANGFELDT J.N. Computer simulation program for evaluation of station keeping system. IFAC/IFIP symposium on Automation in Offshore Oil Field Operation, North Holland, 1976.
4. NORRBIN N.H. Theory and observations on the use of a mathematical model for ship manoeuvring in deep and confined waters. The Swedish Shipbuilding Experimental Tank, Göteborg. Publ. No. 68, 1971.
5. GERRISMA T. et al. The effects of beam on the hydro-dynamic characteristics of ship hulls. Proc. symp. on Naval Hydrodynamics, London, 1974.
6. VAN OORTMERSEN G. and REMERY G.F.M. The mean wave, wind and current forces on offshore structures and their role in the design of mooring systems. Proc. Offshore Technol.Conf., 1973, Houston, 1741.
7. GLANSDORP C.C. Ship tyre modelling for a training simulator. Proc. 4th Ship Control Systems Symposium, Haag, 4, 1975, 117-136.
8. HSU F.H. and BLENKARN K.A. Analysis of peak mooring forces caused by slow vessel drift oscillations in random seas. Proc. Offshore Technol. Conf., 1970, Houston, 1159.

9. SØRHEIM H.-R. Dynamic positioning in single-point mooring - a theoretical analysis of motions, and design and evaluation of an optimal control system. Thesis, University of Trondheim, Report No. 81-105-W.

10. RAINEY R.C.T. The dynamics of tethered platforms. Royal Institution of Naval Architects, 1977, 59-80.

11. BALCHEN J.G. et al. Dynamic positioning using Kalman filtering and optimal control theory. IFAC/IFIP symposium on Automation in Offshore Oil Field Operation, North Holland, 1976.

12. BALCHEN J.G. A dynamic positioning system based on Kalman filtering and optimal control. Modelling, Identification and Control, 1980, Vol. 1, No. 3.

13. SAELID S. A simple parameter tracking filter based on innovation sensitivities. Unpublished work, 1977.

14. JAZWINSKI A.H. Stochastic processes and filtering theory. Academic Press, New York, 1970.

Discussion on Papers 1 and 2

MR B. T. LINFOOT, *Department of Civil Engineering, Heriot-Watt University*

1. I wish to outline the research programme at Heriot-Watt University which involves the development of a number of control systems aiming to reduce peak hawser loads in single-point mooring systems. This work complements the developments reported by the Authors of Paper 2.

2. In our programme, two fundamentally different methods of achieving this objective have been evaluated using instrumented scale models in a wave basin facility (Fig. 1). The first method, similar to that described by Mr Ropstad, involves thruster control of the low-frequency motions of the vessel with the aim of suppressing the low-frequency component of the hawser force record. The second technique involves the deployment of an active tension-compensation device fitted to the mooring winch with the purpose of suppressing the wave-frequency component of the hawser force.

3. Basin trial results shown in Fig. 2 indicate that simple PID-controlled thrusters mounted athwartships at the bow and stern of the vessel are very effective in achieving the desired objective of reducing the low-frequency motions and thereby suppressing the low-frequency component of the mooring load. The wave conditions for this test corresponded to regular waves of 4 m height and period of 6.25 s, with a wind of 1.5 m/s blowing at 45° to the wave direction being applied 300 s into the test. The hawser force record clearly shows the low-frequency fluctuations with the high-frequency loads superimposed. From 150 s into the test the controller effectively suppresses the low-frequency component of the force; the effect of the application of the wind is to cause the vessel to weather-vane round into an equilibrium position

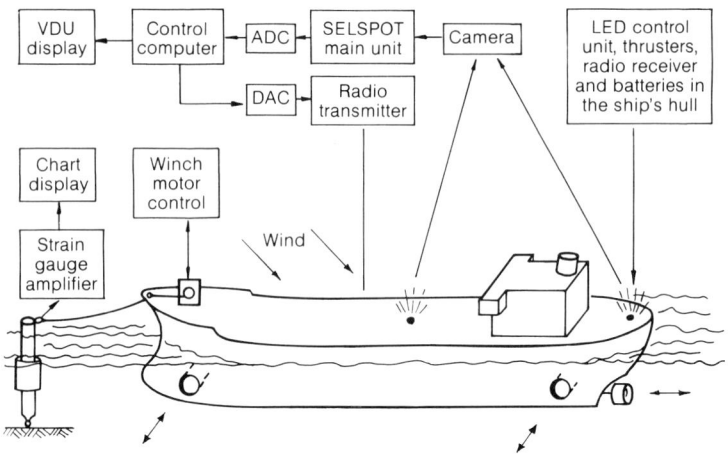

Fig. 1. Schematic diagram of instrumentation

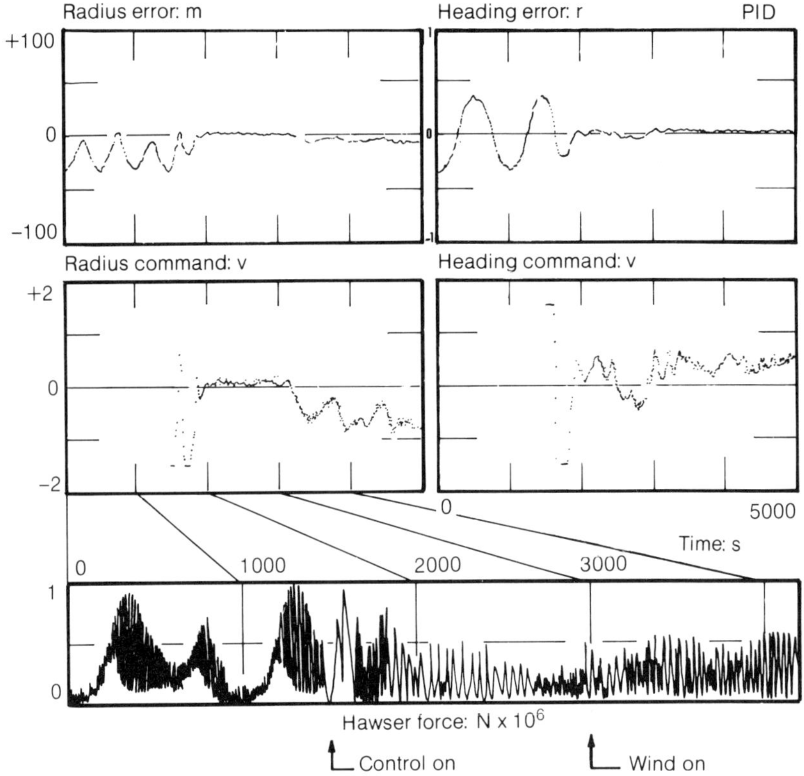

Fig. 2. Regular wave test of motion-controlled vessel

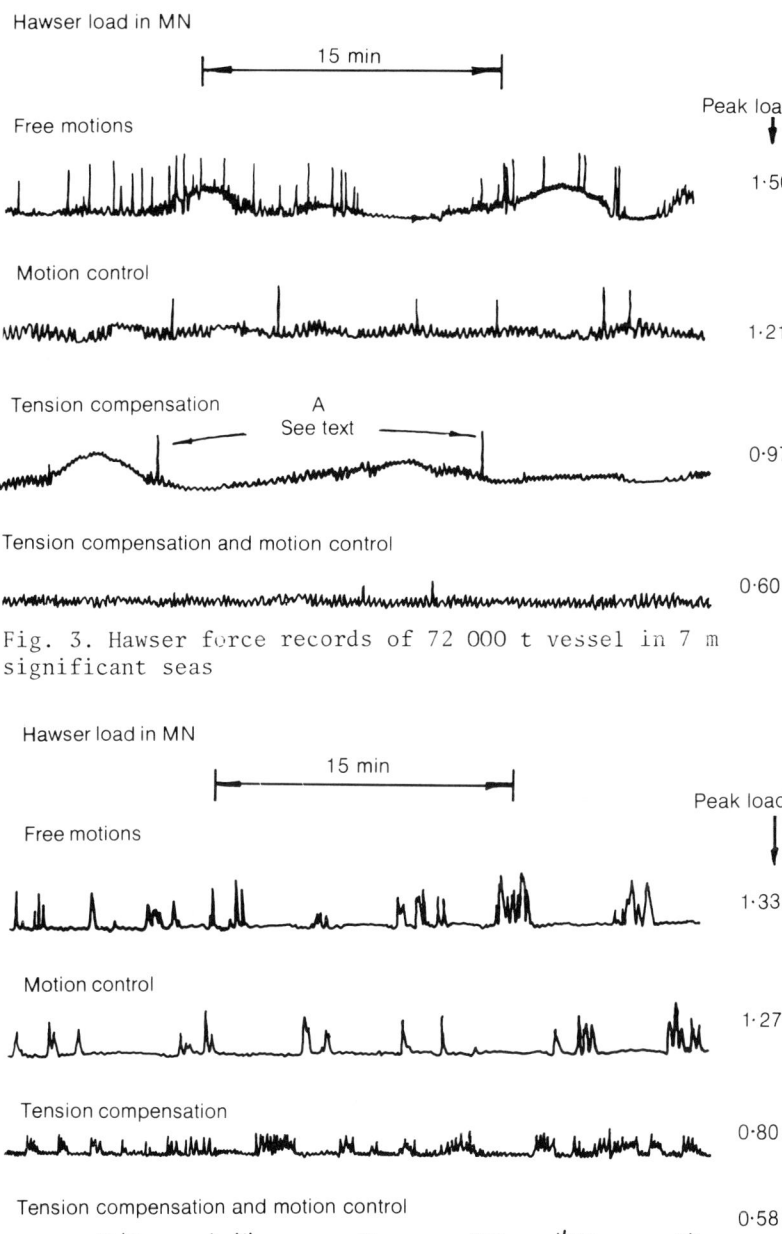

Hawser load in MN

|←——— 15 min ———→|

Peak load

Free motions

1·50

Motion control

1·21

Tension compensation   A
See text

0·97

Tension compensation and motion control

0·60

Fig. 3. Hawser force records of 72 000 t vessel in 7 m significant seas

Hawser load in MN

|←——— 15 min ———→|

Peak load

Free motions

1·33

Motion control

1·27

Tension compensation

0·80

Tension compensation and motion control

0·58

Fig. 4. Hawser force records of tandem moored vessels in 7 m significant seas

at an angle of 20° to the wave direction where the vessel picks up the increased roll motion which is transmitted down the hawser. Fig. 3 shows the results of basin trials with an active tension-compensation device used both alone and in conjunction with a low-frequency motion controller.

4. These tests were conducted on a scale model of a 72 000 t vessel attached to an articulated column buoy by a fairly elastic 70 m hawser. The results, obtained from tests in a scaled 7 m significant sea, show that the tension compensator partially suppresses the high-frequency forces resulting from the first-order responses of the mooring buoy and tanker, and that it is particularly effective when used in conjunction with active thruster control of the low-frequency motions.

5. Typical test results of the combination of thruster control and tension compensation are also presented (Fig. 4) for a 1:125 representation of a 72 000 t tanker moored in tandem to a similar sized yoke-moored storage vessel. Again, the sea state corresponds to a 7 m significant sea and the results show the benefit of using tension compensation in conjunction with active thruster control. Further details of this work are to be found in reference 1.

MR M. W. H. THOMSON, *ICI Petroleum Services Ltd*

6. For a real field an offshore loading system has to be designed relative to the environmental conditions of the location to optimize field economics. This involves weighing the extra cost of a more complex mooring/dynamic stabilization system against benefits such as reduced field storage requirements or reduced field production down time. Such benefits would arise if the improved mooring system convinces the shuttle tanker's captain that he should moor his ship, or keep it moored, in a more severe sea state than would otherwise be the case, so that weather down time for loading operations is reduced.

7. Shuttle tankers may be purpose built, but are more usually conversions of existing ships. The main modifications which could be considered for single-point mooring and loading, apart from the essential minimum are as follows:

(a) a facility to permit slow-astern running of the main engine for long periods
(b) transverse thruster at bow and stern
(c) dynamic winch control of hawser tension
(d) computerized control of thrusters to give automatic

motion stabilization

(e) a forward control bridge to assist ship control while picking up the mooring.

8. There is a law of diminishing returns in buying such 'goodies'. Can the Authors of Paper 2 indicate their relative importance, and the need for fewer or more of these items in relation to the severity of expected environmental conditions?

MR R. W. BREWERTON, *Foster Wheeler Petroleum Development Ltd*

9. With reference to Paper 2, I wish to comment on the application of dynamic positioning (DP) at offshore loading buoys.

10. Regarding the addition of DP and thrusters to tankers for offshore loading, it is not certain whether the increase in conversion cost from about £250 000 to more than ten times that amount justifies the assumed benefits of DP, particularly when the consequent substantial loss of cargo-carrying capacity is considered.

11. The most critical load conditions in the mooring occur when the wind is not aligned with the waves, such as occurs when a cold front passes; under these conditions the tanker rolls considerably. This results in a water flow transverse to the mouth of tunnel thrusters and makes them ineffective. Such a loss of DP effectiveness in marginal sea conditions could result in a sudden large increase in mooring tensions to values that are consistent with the non-DP case. To protect against this situation it may be necessary to ignore the benefits of DP in setting the maximum allowable operating sea state for the unit.

MR ROPSTAD AND DR SØRHEIM, *Paper 2*

12. In the Authors' opinion the following priority exists: item (a) in Mr Thomson's list of modifications, a facility to permit slow-astern operation of the main propeller for long periods of time, is essential; in general terms this feature tends to stabilize low-frequency motions.

13. A transverse thruster at the bow, item (b), increases overall manoeuvrability during pick-up, mooring, and loading. To take full advantage of this feature, control should be at the bow; hence a forward bridge, item (e), is recommended. The benefit of such control has been demonstrated in practice, providing increased safety, reduced hawser excitation and

potential for increased weather limits (obviously the latter
may be a little subjective, since experience so far is rather
limited).

14. With respect to control of low-frequency motions of the
moored tanker, a stern thruster is secondary. The main
potential feature is the weather-vane condition, which is
virtually unaffected by both main- and bow-propeller
operation.

15. Regarding automatic control, either by winch control (c),
and/or dynamic positioning (d), the main features are indicated by Mr Linfoot's comments. Winch control alone reduces
high-frequency tension peaks but has little or no effect on
low-frequency behaviour, including stability. Hence it
should be combined with at least a slow-running astern main
propeller for stability reasons. To the Authors' knowledge,
full-scale trials with automatic winch control have not so
far been conducted.

16. Dynamic positioning (DP) yields far greater potentials.
With both bow and stern lateral thrusters, and a controllable
(pitch) main propeller, complete low-frequency motion control
is achieved. In principle, hawser tension may be reduced and
kept at zero tension level (no hawser required except for
safety/back-up reasons). However, the main features of DP
(low-frequency control) are gained without a lateral stern
thrust.[2] Hence, since manoeuvrability requirements more generally may require both controllable-pitch propeller and bow
thruster, the inclusion of DP does not necessarily involve
extensive construction/conversion of the hull and machinery.

17. Experience so far indicates that once ship captains gain
confidence in the system, a major advantage of DP is the
reduced human strain. Keeping a loading operation going
during a 10 h loading period, even in moderate weather, is
a tough job. Any equipment which can add significantly to
this process should therefore be considered cost effective.

18. Operationally, hawser wear is reduced and operations may
be extended for longer periods and into more severe
conditions (quantitative assessment is clearly impossible at
this stage). On the negative side increased fuel consumption
and propeller wear by continuous operation are acknowledged.

19. The high-frequency hawser peaks which are left during DP
operation are normally at a very moderate level and fully
acceptable. Hence the additional features of DP favour this
solution when automatic (dynamic) control is desirable.

20. In reply to Mr Brewerton, the Authors agree completely with him in relation to the existence and potential effect of rolling in certain sea and wind/current conditions. However, in some situations or within a reasonably short time, the problem can be greatly reduced.

21. Obviously the problem may only occur when the ship is high in the water, i.e. during ballast. Some tankers have segregated ballast tanks, which then make it possible to keep ballast condition at a larger draft and exchange ballast and crude at the initial phase of loading. Secondly, tanks in the foremost region of the hull are filled up first, giving the ship a temporarily forward trim which lowers the important forward tunnel into the water.

22. If the DP system includes a lateral stern thruster, this can be used to weather-vane the ship to a more favourable heading, giving less rolling motion. Depending on weather condition, ship, and thruster size, some 10-30° change in heading can be achieved.

REFERENCES

1. LINFOOT B. T. et al. Control of the low-frequency motions of single-point moored vessels. OTC 4349, Offshore Technology Conference, Houston, USA, 3-6 May 1982.

2. SØRHEIM H.-R. Dynamic positioning in single-point moorings - a theoretical analysis of motions, and design and evaluation of an optimal control system. Dr Ing. thesis, University of Trondheim, 1981.

3　Moored production and loading facilities as experienced at the Argyll Field

R. J. COCKRILL, Hamilton Brothers Oil and Gas Ltd

SYNOPSIS. This paper is presented as a narrative description of the facilities installed at the Argyll Field. The major problems encountered with this unique installation are discussed, along with the authors opinion as to possible improvements which may be considered if this type of facility were to be repeated.

INTRODUCTION

1. There are many types of moored/floating production facilities either in existence or conceptual, ranging from tethered tankers serving one well completions through conventionally moored semi submersible barges to tension leg platforms. For the purposes of this paper it is intended to restrict the discussion to within the author's experience, as of the end of 1981, at the North Sea, Block 30/24, Argyll Field (Ref. 1) where the first ever floating production facility is still in service.

2. No floating production facility could ever be conceived without consideration of the production system as a whole, that is the producing/injection wellheads through the facility to the product exporting route, whether this be pipeline or tanker. Therefore, although each of these portions of the production system are worthy of papers in their own right, it is intended to attempt to cover the whole concept, albeit rather briefly.

3. The sequence of addressing the components making up the whole facility will be that as followed by the oil process; wellhead, production riser, separation onboard the main floating barge through to the export tanker loading system.

SUBSEA WELLHEADS

4. The Argyll Field lies in Block 30/24, 190 miles south
east of Aberdeen and 25 miles south west of the Ekofisk
Field (see Fig.1). The Field was discovered in 1971 and
brought to production in June, 1975. The main hydrocarbon
bearing structure is a Zechstein dolomite overlaying a
Rotlegendes sandstone which is produced via subsea completed wells using 'wet' Christmas Trees positioned as satel-
lites to the floating facility. The Christmas Trees are
very simple in construction, controlled by hydraulic fluid
and consisting in the main of manual and hydraulic master
valves, hydraulic production wing valve and hydraulic swab
valve. There is also a hydraulic annulus valve tied back
into the production flowline (see Fig.2).

5. This description is of the base unit which makes up
the majority of the producing wells, and can be enlarged
upon by the addition of a hydraulic service wing valve for
the eventuality of tying more than one well into flowline
system in order to give the facility to test each well in
isolation from the others (see Fig.3). This has not been
necessary with the wells producing to date, although
Hamilton Brothers have completed one well on a remote structure and another on a closer structure that can be tied into
the same flowline. These Christmas Trees do have the dual
wing valve configuration to allow for expansion of both or
either of these structures by further drilling should the
reserves prove commercial and producable. To interconnect
wells in this fashion, a flowline of dual construction must
be laid back to the floating facility.

6. Control of these wellheads is by simple hydraulic
fluid pressure, one signal per function, via a multicored
umbilical laid onto the seabed.

7. Wells within approximately two miles of the moored
facility do not require hydraulic accumulators on their
Christmas Trees to maintain control. Hence with the exception of the two remote wells mentioned above there are no
subsea hydraulic control accumulators, instead the main
system utilises three-way 'dump' valves installed directly
onto each valve actuator. The 'dump' valves function by
allowing pressure from the umbilical to enter the actuator,
thus effecting the operation, upon release of the fluid
pressure the 'dump' valve shuttles to open the third port
and expel the actuator fluid to the sea. This enables an
insignificantly delayed valve-opening and an immediate fail
safe closing.

8. The downhole safety valves are controlled by direct
fluid pressure which is expelled back at the control panel

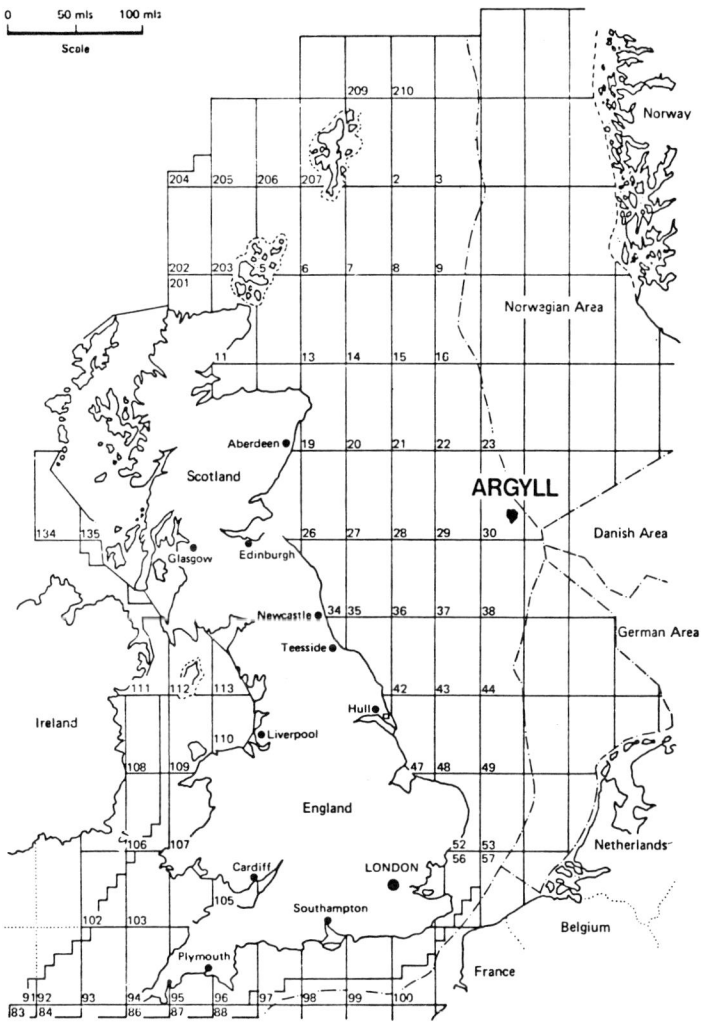

Fig. 1. The Argyll Field

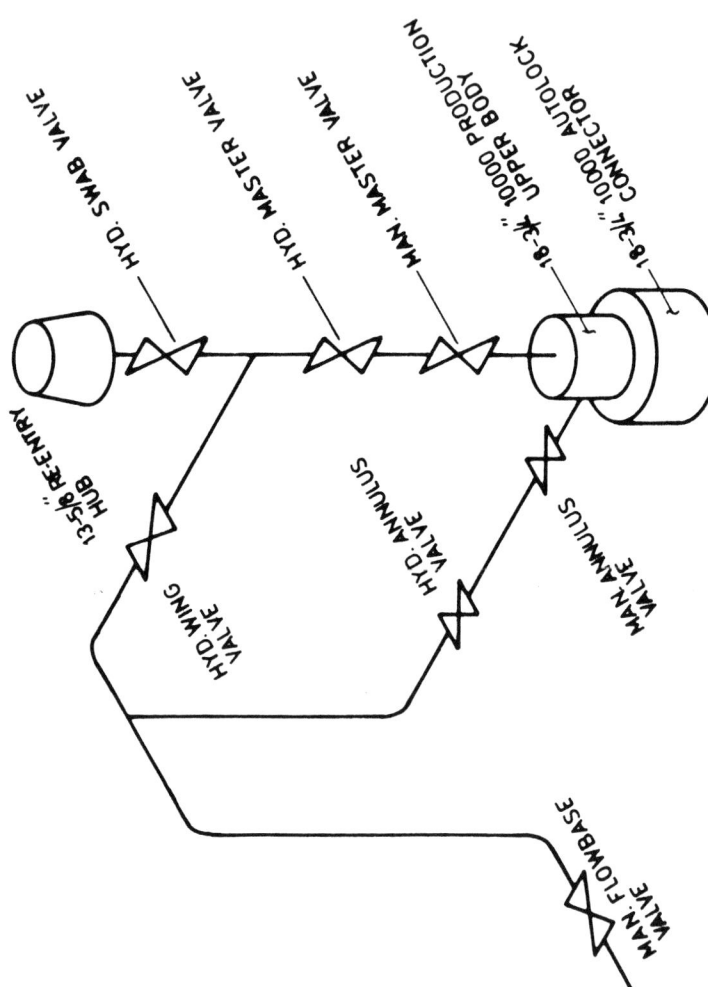

Fig. 2. Schematic of sub-sea tree (Argyll)

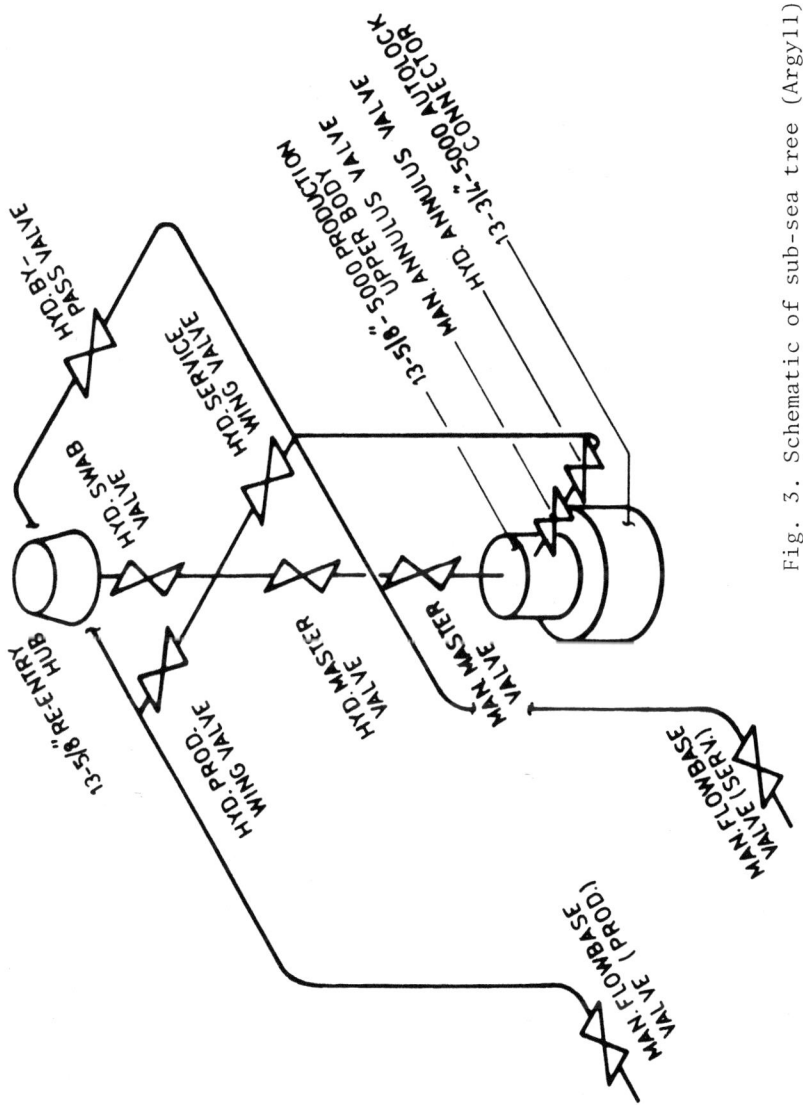

Fig. 3. Schematic of sub-sea tree (Argyll)

onboard the production facility.

9. The system generally operates very satisfactorily, all problems arising are because of the remoteness of the equipment from the main facility. Major problems experienced are:- downhole safety valve failures and control system conduit failures.

10. The downhole safety valve failures require the mobilisation of a semi-submersible drilling rig to carry out either a wire-line replacement of the faulty valve or installation of a wire-line retrievable valve into a tubing mounted valve and seating nipple. In spite of this the Argyll operation probably experiences a better than average life expectancy from its downhole safety valves.

11. The control system failures are mainly caused by leaks or hoses becoming disconnected because of corrosion of the coupling. The latter problem has now been cured by complete replacement of all couplings with stainless steel ones.

12. Remoteness also makes fault diagnoses extremely difficult, for instance, if a downhole safety valve fails to open, or suddenly closes, with massive fluid loss via that function, then there are two possible causes, valve failure, usually the valve-to-tubing packing, or control hose failure. What usually happens in this event is that divers from the main floating facility will establish the integrity of that suspect function at the furthest accessible point. If that function is intact then all that this has told you is that there is still an unknown problem downstream of that point. A diving support vessel would be called-in to repeat the above exercise at the wellhead by simply closing a half-inch valve. A pressure test of the function confirms if the problem is downhole or on the seabed. A downhole problem requires a workover by a drilling rig, a subsea problem needs the continued use of the diving support vessel.

13. Improvements to this system are simple, template directional drilling of the wells which is now reasonably commonplace would provide immense benefits. Easy access to well tubing for workover by the floating production facility, easy access to the Christmas Trees. Maintenance would be simpler and pay dividends by reducing lost production, negate the need for a diving support vessel and make flowlines and long control umbilicals redundant. The capital saved by not installing flowlines usually compensates for the directional drilling costs.

14. A typical Argyll subsea wellhead and Christmas tree assembly is shown in Fig.4.

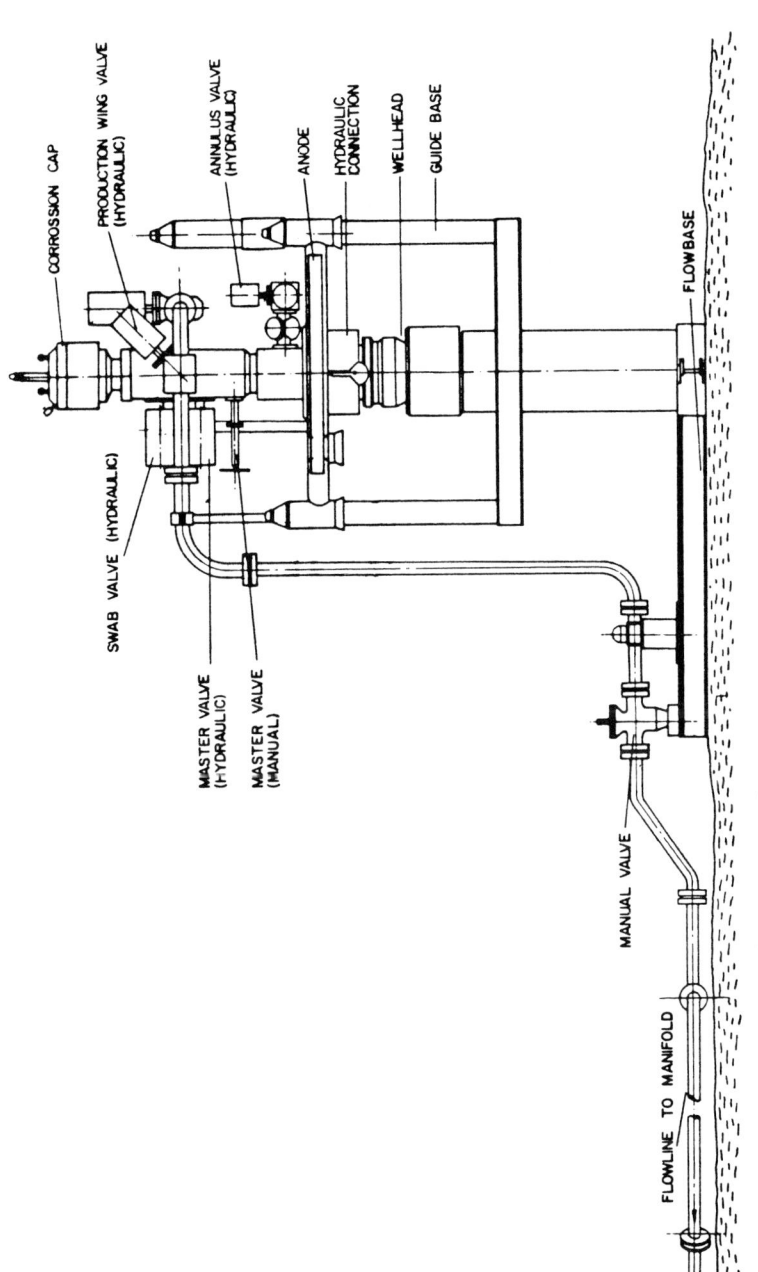

CORROSSION CAP

PRODUCTION WING VALVE
(HYDRAULIC)

ANNULUS VALVE
(HYDRAULIC)

ANODE

HYDRAULIC
CONNECTION

WELLHEAD

GUIDE BASE

FLOWBASE

SWAB VALVE (HYDRAULIC)

MASTER VALVE
(HYDRAULIC)

MASTER VALVE
(MANUAL)

MANUAL VALVE

FLOWLINE TO MANIFOLD

Fig. 4. Typical Argyll sub-sea well head and Christmas Tree

PRODUCTION RISER SYSTEM

15. The Production Riser System at Argyll is an assembly of standard drilling components brought together, in what was a unique manner in 1975. The system consists of five basic elements from the seabed upwards, the mass anchor, permanent base, manifold, risers and flexible connections (see Fig.5).

16. The mass anchor is precisely what its name implies, it is a concrete base 40 ft. x 40 ft. x 500 tons with a steel frame, vertical guide posts and a central vertical 42 in. diameter conductor pipe. The conductor pipe is used as a support and latching member for the permanent base, the guide posts assist in locating the permanent base. The sides of the mass anchor slope downwards towards the seabed and contain saddles for restraining the incoming flowlines.

17. The permanent base is an extension of the drilling concept permanent guide base, complete with guide wires from it's four corner guide posts. The guide base concept has been expanded to provide pipework for interconnection of the subsea flowlines with the next element, the riser manifold. Incorporated within the permanent base are manually operated valves for closing off each flowline should the need arise. These are diver operated.

18. Stabbing into the permanent base is the riser manifold, as expected it is located by the base guide-posts. The manifold is latched into position hydraulically. Incoming wells are connected from base to manifold by parallel male to female "choke and kill" connectors, the outgoing export lines connect by hydraulic connections which also serve to secure the manifold. The pipework within the manifold is arranged to connect each well to it's individual riser via a hydraulically actuated valve. Each well is also able to be routed, via a valved tee, into a "ring main" facility. This ring main serves two functions, one is to enable each well to be produced via any riser, for riser maintenance, and in order that each riser can be used to service any other riser for washout purposes. The manifold pipework is terminated on top of the manifold with hydraulically operated connectors. Risers which stab into these connectors are arranged radially around a larger, central riser.

19. The central riser is 10 in. nominal bore and serves as the main supporting member and export or shipping riser. This member consists of a stab sub-assembly, universal joint and 6 x 40 ft. joints of 10 in. riser pipe connected with standard marine riser joints. There are 8 production risers positioned radially around the 10 in. riser, each 4 in.N.B., made up of a stab sub, safety joint, one 6 in. x 40 ft.

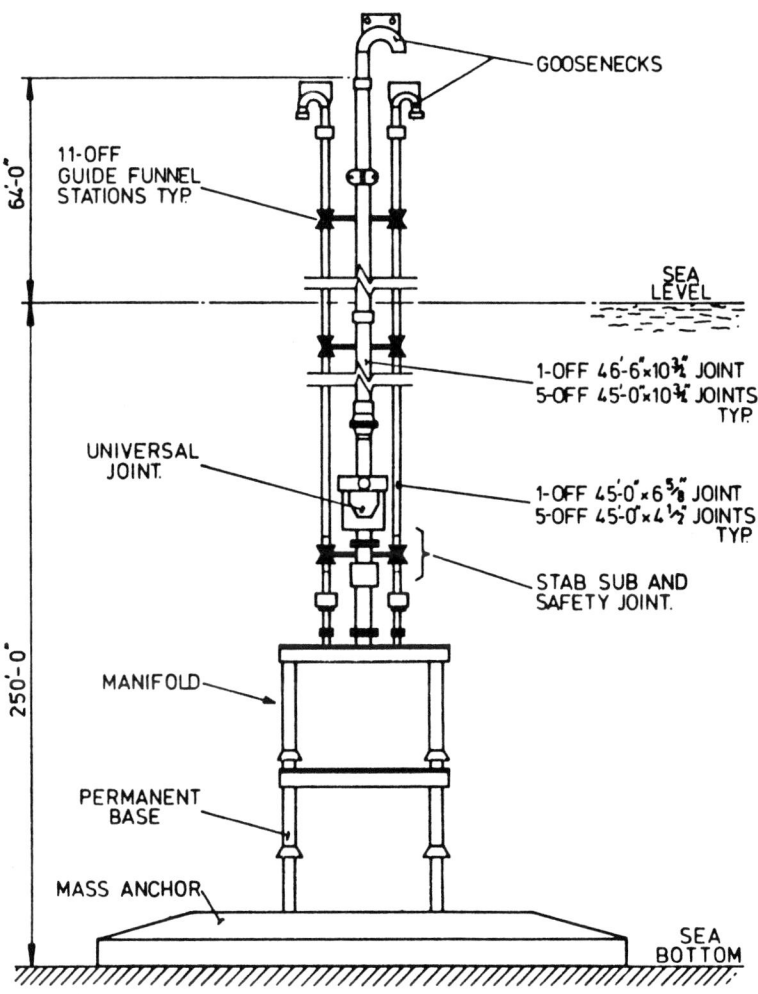

Fig. 5. Argyll production riser system

riser joint and 5 x 40 ft. joints of 4 in. riser pipe connected by $4\frac{1}{2}$ in. I.F. tool joints. Lateral movement of the
floating production facility is transmitted to the top of
the riser assembly which is also fixed at the seabed, hence
deflects to accommodate the movement. The 10 in. riser
accommodates this movement via the universal joint, the
4 in. risers bend to follow this deflection, hence the larger diameter greater wall thickness bottom pipe joint, which
contributes greater strength to tolerate the continuous
flexing forces. Positioning of the 4 in. risers is maintained by guide funnels located radially and at regular intervals over the length of the 10 in. riser. The radius of
bending of the 4 in. risers around the 10 in. universal
joint is maximised by the absence of guide funnels over some
98 ft. of the lower ends of the riser assembly.

20. Goosenecks terminate the upper ends of each of the
risers in the barge moonpool, these connect to the fixed
pipework of the production facilities via flexible pipes.
Standard drilling riser tensioning units are used to support
the assembly and maintain it in tension to eliminate destructive cyclic stresses.

21. As can be imagined steel riser systems have inherent
disadvantages, mainly weather sensitivity which causes loss
of production during bad weather periods because of movement
of the barge. Barge movement is a complex subject, it consists of four main elements occurring simultaneously. These
elements are heave, vertical movement; roll, port to star-
board movement; pitch, forward to aft movement and excursion, horizontal displacement from location. A riser remaining connected during severe weather situations is liable
to damage by vertical collision with the barge drill-floor
or deck and by horizontal collision with the barge deck or
hull structure subsea. Hence it must be disconnected, pulled and stood back in the derrick. The result of this is
production shutdown and lost revenue.

22. Occasionally damage can be done to the manifold pipework by loosening of flanged joints caused by alternating
flexing stresses transmitted via the riser during marginal
weather conditions. The use of backing lock-nuts has been
found to be a simple and effective cure of this problem.

23. Riser handling equipment takes the form of a traditional drilling derrick, complete with draw works and accom-
panying equipment. This is not advantageous to the operation of a floating production unit because it adds to the
unit's instability during bad weather conditions by raising
it's centre of gravity. Riser fatigue control is carried

out on an annual basis by shipping the riser ashore for a comprehensive N.D.T. inspection. This inspection usually takes the form of a critical visual inspection for obvious defects, a full body M.P.I. of all riser pipes, for detection of surface cracks, X-ray and ultrasonic scanning of all butt welds for cracking of the heat effected zones or weld metal, and ultrasonic wall thickness checks to detect any erosion damage. In addition to this the tool joints are specially surveyed for tong damage, thread damage and seal area damage. Repairs being carried out as necessary.

24. These inspections are closely monitored and sometimes witnessed by the Department of Energy, Pipeline Engineering Division.

25. Possible improvements to this system are the installation of a "singly-fixed" manifold and flexible risers. It is considered that installation of the subsea manifold in the same manner as B.O.P. stacks would be suitable, this would result in good harmony with, say, a template wellhead assembly, utilising an identical hydraulic connector.

26. Although the author has no experience of flexible risers they are in use in other locations, for instance the Enchova Field offshore Brazil, with much success. The possible advantages of using this type of riser seem to be the lower sensitivity to weather, less fatigue damage, less handling equipment; all of which result in less lost production caused by downtime.

27. If a steel riser system is selected or imposed upon a design team then perhaps a variation on the theme is worth considering as a possible improvement. This variation is to site the choke manifold onto the seabed. A considerable advantage of doing this is to reduce the number of risers from, say, 9 in the case of Argyll, to only 3. That is a reduction from one incoming riser per well, plus the export riser, to two incoming risers plus the export riser. The incoming risers will then be one collective feeder for the processing plant, carrying the co-mingled fluids from each well; and the test riser, into which each well could be separated for testing purposes. A definite disadvantage of this variation is the need to place more equipment onto the seabed and increase the number of control functions by at least four per well, probably more, depending upon the amount of feedback data required. However, there are less risers to handle. This will not be the case if secondary recovery methods are to be employed, but the overall number of risers will still be reduced.

FLOATING PRODUCTION FACILITIES

28. The facility in use at the Argyll Field is the semi-
submersible barge Transworld 58 (TW58), which was converted
from a drilling rig for it's present use in 1975. The TW58
is moored on twelve chain cable connected anchors, each with
piggy-back anchors. The derrick and draw-works remain onboard for riser handling.

29. The production equipment as installed by Hamilton
Brothers was designed to handle 70,000 BOPD, stabilising the
crude oil to be suitable for shipment by sea tankers. The
facilities are very simple, consisting of a choke manifold,
two stages of three-phase gas-oil-and-water separation, and
an atmospheric surge tank. There is also a test separator
facility placed in parallel to the first stage separator.
From the surge tank the stabilised crude oil is pumped via a
volume metering system to the 10 in. export riser (see
Fig.6). An additional piece of equipment was added to the
installation in 1978, shortly after the Field began to produce water. This is a Wemco Depurator, which is an intrain-
ed gas floatation oily water separator from which cleaned
water is dumped directly to the sea. Gas flaring is via two
horizontally deployed booms, each capable of flaring up to
44 MMSCF/D of gas and 10,000 BOPD crude oil.

30. In addition to the production facilities, Hamilton
Brothers have installed, through a contractor, a 1,000 ft. 6
man saturation diving spread. This is in order to enable
maintenance to be carried out on the subsea manifold.

31. There are no gas lift, nor secondary recovery facilities onboard the TW58, in fact there is considerably doubt
as to the feasibility of being able to install them. The
limiting factors for installing secondary recovery equipment
would be deck space and deck loading capability. Happily
this does not present a problem at Argyll because of the
natural aquifer drive system present in the reservoir. In
this respect Argyll is ideally suited for floating production exploitation using the satellite completed well system,
for had there been problems with sand, wax or a severe corrosion problem, then it may have been a completely different
story.

32. Generally the production system copes very well with
one notable exception, that being the water separation.

33. Because of the confined deck space and deck loading
available the residence time for gas, oil and water separation is in the order of 15 minutes. During this limited
period only minimal gravity separation of water can be
achieved, hence this is assisted chemically, using emulsion

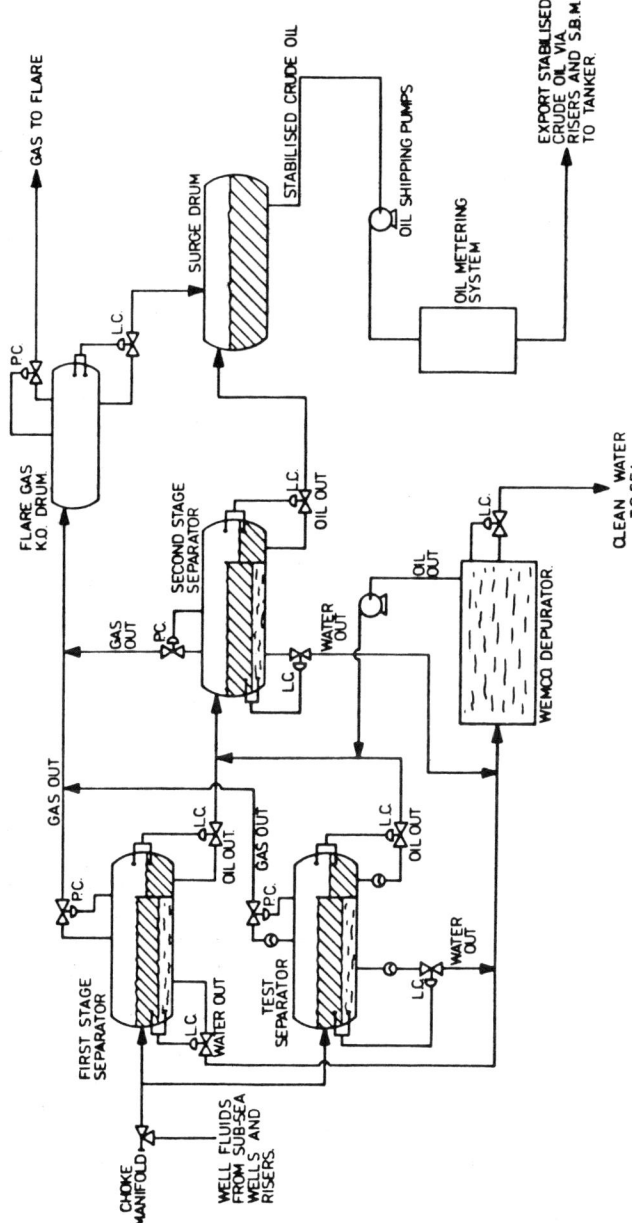

Fig. 6. Schematic of Argyll production system

breaking surfactants. The chemical surfactant, is injected
into the production stream from each well, individually, as
it leaves it's production riser and prior to being sheared
by passage through the choke. The emulsion breaker serves
two purposes, it accelerates the water separation and segre-
gation and clarifies the oil-water interface in the production separators. Good interface clarity is essential for
good quality crude oil and clean water effluent, particularly when producing from a floating production facility.
Cloudy, ill-defined interfaces do not allow for proper water
level control, which in conjunction with the slopping caused
by the barge movement results in either an export crude oil
with a high water content, or an effluent water quality
which is unacceptably high in residual oil.

34. At Argyll an export crude oil B.S. & W. of less than
0.5% is easily attainable, at times this has been maintained
lower than 0.1% although this is sometimes sacrificed for
improved effluent water quality.

35. Since the installation of the Wemco Depurator the effluent water quality has been maintained well below the imposed limit of 75 ppm, set by the Department of Energy. In
fact on good calm weather days it can be as low as 10 to 15
ppm rising to 50 to 60 ppm during marginal weather states.
Again the main problem is level control. The Wemco is a
tank divided into cells with sealing lids. The oily water
passes through the unit cell by cell. A pump takes clean
water from the last cell and recycles it back into the
others via cover penetrations which extend to the bottom of
each cell. Beneath the cover, at each penetration, is a
venturi which pulls in gas from the void space between the
liquid level and the cover. The gas is that which is released from the formation water. As the gas bubbles back to
the liquid surface it takes with it the oil contained in the
water. The oil accumulates on the water surface and is
skimmed off by rotating paddles to be recycled through the
separation plant. A chemical coagulant is injected into the
Wemco inlet water stream to assist with the floatation process.

36. Generally speaking there are two types of problems
arising from the TW58 type of facility, those caused by the
conversion of an existing drilling rig to a production unit,
and those caused by the limitations of using a floating
barge. However, there are also definite advantages.

37. Conversion leads to compromise, which is never an ideal
situation. For instance a drilling rig designed hydrodynamically for a compromise between on-location stability, to

optimise drilling time, and "towability" for minimising
drilling time lost during location changes. A floating production unit does not require to be towed, excepting for removal to dry-dock or sheltered water for remedial work, but
does need to be exceptionally stable on location to be able
to maximise producing time. Converted drilling rigs do not
have the deck space and loading capability to accommodate
the large, heavy equipment required for the more usual type
of oilfield which needs application of secondary recovery
techniques.

38. Apart from these obvious problems there are economic
reasons why it is usually impractical to convert drilling
rigs to production units, particularly at the present time.
The demand and prices attracted by drilling barges is so
high at present that to purchase a barge for conversion
would probably be prohibitive, hence day-rate charter is the
alternative. Charter parties on converted barges can be
very costly, not only is the day-rate high, but it would
probably be a contractual requirement to re-convert the rig
back to the drilling mode after it's usefulness as a floating production facility has expired.

39. A limitation which has not been mentioned is that of
water depth, both shallow water and deep water. At present
the workable water depth for a floating production unit is
between 100 to 600 ft. The shallow water presenting difficulties with the barge hull clearance of the seabed located
facilities, remembering that most semi-submersibles have
operating drafts around 75 ft. The deep water limitation is
being caused by the barge mooring and station keeping capability combined with riser design and handling, and subsea
maintenance problems. The latter is able to be overcome by
appropriate design at the conceptual stages of development,
but would probably result in a purpose built facility, such
as a Tension Leg Platform, which is outside the scope of
this paper.

40. In certain circumstances, the advantages of using a
floating production facility can outweigh the disadvantages.
For example, a company making a discovery could temporarily
mobilise a floating unit for two very sound reasons. Firstly to conduct a prolonged production test of the discovered
structure for data collection reasons, as a sound basis for
designing and constructing a larger, probably piled jacket,
installation. Secondly, to establish a source of revenue at
a time when resources are stretched by the financing of
exploration and construction programmes. Of course some
discoveries are just not worthy of a permanent elaborate

facility, in which case the floating unit is a good alternative on such structures. These structures could be isolated pockets of hydrocarbons or smaller ones adjacent to an
established oil or gas field.

TANKER LOADING

41. Floating production facilities lend themselves to use
for exporting stabilised crude oil via pipelines or tanker
loading facilities. Because of weather sensitivity, tanker
loading is less advantageous than pipeline export, but this
is an economic criteria, not simply a decision of preference. The method of exporting crude at the Argyll Field is
the tanker loading method.

42. There are many different classes of tanker mooring and
loading facilities ranging from "piggy-back" tanker to tanker operations, through loading buoys to large gravity based
tower structures. At Argyll the loading is via a catenary-
anchor-legged-mooring (CALM) buoy of Single Buoy Mooring
(SBM) manufacture (see Fig.7). The buoy is located 1.4
miles east of the TW58 and connected by a 10 in. nominal
bore flowline, fed by the 10 in. central riser.

43. The hull of the buoy takes the shape of a cylindrical
tank 36 ft. diameter and 17 ft. deep, with a hollow centre
12 ft. in diameter. This has a draft of 12 ft. 9 Ft. below
sea level a skirt protrudes which contains six stoppered
apertures for the chain cable moorings. The hollow centre
of the buoy acts as a "moon-pool" through which the production riser passes. On top of the hull a 13 ft. diameter
slew ring type bearing supports a turntable assembly which
provides both an anchorage for the tanker mooring hawser and
support for the tanker loading pipework.

44. Mooring of the SBM is by means of six 3-5/16 in. stud-
link-chain-cable catenaries attached to 15 ton anchors.
Each anchor is situated some 2,500 ft. from the buoy and is
reinforced by a length of ground chain and piggyback anchor.
The chains are pretensioned to approximately 50 tons, which
is monitored by chain angle readings taken immediately
beneath the skirt of the buoy, at the top of each chain's
catenary.

45. A tanker mooring hawser connection is made to the turntable via plate-and-pin construction 'Y' shaped yoke. The
arms of the yoke are secured to the turntable with the end
of the leg terminating in a retractable pin assembly. The
yoke is able to move vertically as the tanker rides on the
mooring, being cushioned from battering the turntable by a
block of rubber. Hawser attachment to the pin is made by a

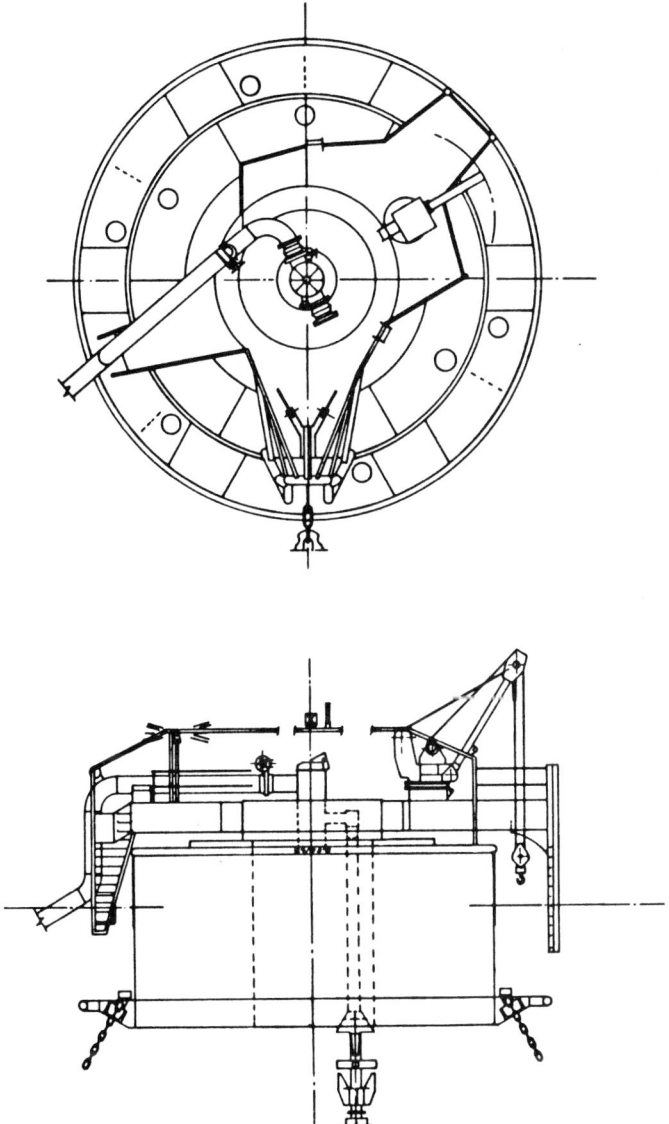

Fig. 7. SBM tanker loading buoy as installed at the Argyll Field

plate link and shackle assembly. Mooring hawsers are many
and varied, being constructed from different types of man-
made fibre, plaited in different manners. At Argyll we use
a polypropylene fibre hawser constructed as a grommet, that
is a continuous loop of 21 in. circumference hawser made up
through two thimbles, resulting in a double legged hawser
when the thimbles are extended. Attachment of the tanker to
the end of the hawser is by a short length of 3 in. stud
link chain cable, known as a chafe chain, which passes over
a fairleader in the tanker's bow and is made fast to a hook
located on the fo'c'sle.

46. Production pipework connection is made from the 10 in.
pipeline on the seabed to the SBM via a 12 in. hose assembly. The hose is connected to a pipeline-end-module (PLEM)
mass anchor, and is made up of 10 joints of 12 in. diameter
x 40 ft. long hose. Two joints of hose (80 ft.) distant
from the seabed is connected a buoyancy tank, the purpose
of which is to maintain the lower section of hose in tension
and off the seabed. On leaving the buoyancy tank the hose
takes the shape of a catenary before rising to the buoy,
this provides slack in the hose string to accommodate vertical movement and excursion of the unit. The buoy connection
is made via a universal joint situated immediately beneath
it, and a riser pipe which is brought up through the moon-
pool (see Fig.8). This riser pipe is secured to the hull of
the SBM by a swivelling collar which allows for rotation
during the excursion of the buoy. A flexible rubber joint
finalises the connection to the SBM's production pipework.

47. Production pipework is 20 in. diameter and passes over
the turntable. In order to allow for the turntable's rotation a stuffing box arrangement is utilised to make the connection between the moon-pool and the turntable. Dropping
over the side of the buoy's turntable to sea level the tanker loading hose connection is made to the buoy. The loading hose is manufactured in 35 ft. lengths with buoyancy to
enable it to float. It is tapered from 20 in. diameter to
6 in. diameter. The tanker end connection is made vertically over the side of the vessel's bow, utilising a hydraulic clamp.

48. Argyll has two dedicated tankers which operate off the
loading facility, the 'Leonidas' which is 43,000 D.W.T. and
the 'Spiros' which is 46,500 D.W.T. although the facility is
suitable to accept tankers of up to 100,000 D.W.T.

49. Utilities onboard the SBM consist of a 30 ton hydraulic
crane and tugger winch, each driven by an in-situ hydraulic
power pack or a skid mounted power pack from a nearby work

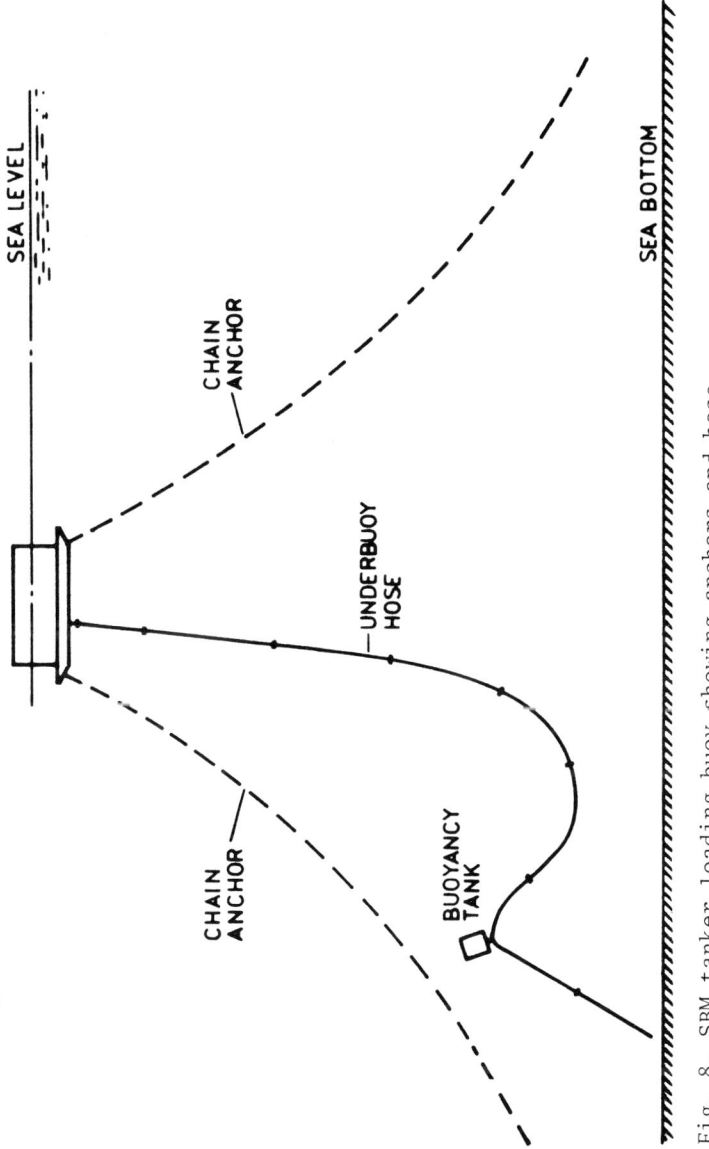

Fig. 8. SBM tanker loading buoy showing anchors and hose

boat. A third winch, air driven, is also available, this is powered by utility air also taken from a work boat when moored alongside the buoy during maintenance and repair. Navigation aids in the form of lights and a foghorn are also installed.

50. Experience at Argyll has proven that this type of tanker loading is a maintenance intensive and weather sensitive operation.

51. Although at face value the moorings of the loading buoy are more than ample when compared with infinitely larger structures such as semi-submersible drilling barges, which use 8 anchors on 3 in. chain cable. The maintenance problems are of a different nature. In the dynamic state a loading buoy has to be considered from it's mooring anchors, through their chains to the buoy, then through the hawser to the tanker. When tension is applied to the hawser by the wind and sea effect on the tanker forces are applied which displace the buoy in the direction of the force, stretch the hawser, and lengthen the mooring chain catenary by picking up chain off the seabed. When the tension decreases the chain is laid back down and the hawser shortens, as the system returns to normal. This cyclic action is continuous and varies in severity according to the weather conditions. It has two effects, one is for the anchor chains to wear out as they "thrash" up and down on the seabed, the other is for the mooring hawser to fatigue.

52. As a consequence of this the mooring chains are inspected, subsea, annually and the thrash zone shots of chain replaced bi-annually. The tanker mooring is changed twice yearly, unless other damage occurs, in which case it is changed at once.

53. Drilling barge anchor chains are maintained in much greater tension, hence limiting the amount of thrash. The barge, being a much greater mass, is less effected by weather conditions in terms of displacement from location, also less excursion can be tolerated during drilling operations, hence minimal chain movement is experienced. This smaller movement of chain results in less wear, which in any event is more easily inspected and corrected as anchors are pulled at frequent intervals during rig moves.

54. As can be imagined the loading facility described is very weather sensitive, the effect being that production is interrupted whenever loading has to cease.

55. Maintenance and repair of this seemingly simple installation is a major operation at Argyll, maintenance being mainly lubrication and inspection by nature and repair being

very intensive. In general, maintenance takes the form of greasing the turntable bearing, draining the bearing labyrinth of seawater, inspection for damage and wear, and monitoring the buoy mooring chain tensions by measuring their angles as mentioned above.

54. Repairs commonly carried out on location are replacement of the floating hose string, the underbuoy hose string and mooring hawser, and general minor structural repairs, although as stated above hose and hawser replacement is done periodically as part of the planned maintenance scheme. Damage to hoses and hawser can be sustained either by the elements or by the tankers and work boats. Repairs to damaged buoy utilities are also required occasionally, as often this equipment is swept away by the elements.

57. As is well known in the industry the turntable bearing on the Argyll SBM has been replaced on two occasions. The first time this was effected by towing the buoy to Rotterdam and putting it in a slipway for the repair. The second time a heavy lift barge was used and the buoy repaired on it's deck. It is necessary for the buoy to be maintained still and level whilst the bearing is set on "chock-fast" resin, hence during the latter repair the barge was sailed to a sheltered anchorage in the Firth of Forth.

58. Because of it's close proximity to the sea, production losses caused by loading downtime are exaggerated by waiting for suitable weather conditions to effect the repair. As a rule of thumb, sea states of less than 8 ft. are required before safe access can be made to the buoy, even then working conditions are extremely difficult.

59. As a result of this difficulty, many modifications have been made to the Argyll loading buoy in order to simplify maintenance and repair work. Two examples of these modifications being a hydraulic mooring hawser pin extractor, and the raising of the floating hose to buoy connection, with installation of a davit, to prevent the connection having to be made in the surf zone.

60. Experience has shown that, as with most mechanical equipment, frequent inspections of a loading buoy are essential to maximise loading time. It is often found that an inspection will reveal defects before they become serious, hence allowing for planned repair or replacement during a routine shutdown. For example, superficial damage to the floatation of a loading hose would soon become a major problem with a risk of pollution if it were left without being replaced. This type of damage does not effect the integrity of a hose unless it is neglected and left to the sea to wor-

sen the damage by corroding the reinforcing steel of the
hose. Our policy in a case like this is to change out the
hose at the next opportunity, such as tanker change, or
after a natural break in loading caused by bad weather.
61. A schematic illustration of the Argyll Field is shown
in Fig.9, with photographs of the Transworld 58 and SBM
mooring buoy and tanker in Figs.10 and 11 respectively.

OPERATING PARAMETERS
62. The operating parameters of the Argyll Field have been
continuously updated by experience and consultation with
Certifying Authorities and Insurance Underwriters. The
limits upon the use of floating systems are imposed by the
weather. Our philosophy is to react to it's effect, not the
weather itself. It must be stressed at this point that the
guidelines for reacting to bad weather do not preclude the
use of experience and discretion by the people operating the
facilities. Generally speaking a bad weather pattern would
result in the wind and sea state building gradually, in
which case normal procedures would be put into action.
Sometimes, the weather changes rapidly calling for early
action to secure the installation before it hits. In this
case a good knowledge of the weather and sea is essential,
as is a reliable weather forecasting service.
63. With the production riser installed, connecting the
Transworld 58 to the subsea manifold, and a tanker moored to
the SBM and loading normally a typically weather related
field shutdown would proceed as follows:
64. The weather would worsen, usually with the wind increasing from north west, causing the seas to build. The effect
of this being to slowly increase the tanker mooring hawser
tension. An operating limit of 100 tons tension is imposed
on the tanker loading, this tension is monitored and recorded both on the focsle and bridge of the tanker. Conditions
prevailing at 100 tons hawser tension would approximate to
40-45 kt. winds and 25-30 ft. seas, although weather direction, wave period and duration and state of loading causes
this to vary considerably.
65. When the tanker mooring hawser tension reaches about
85 tons, production at Argyll is shut down and the riser
system displaced with water. All production separator
fluid is emptied, oil to the tanker and water to the sea,
the export riser is then displaced with water down to the
manifold. All hydraulic valves are closed at the manifold
and the master and wing valves at the wellheads. The tanker
is advised that he can leave the mooring. Production

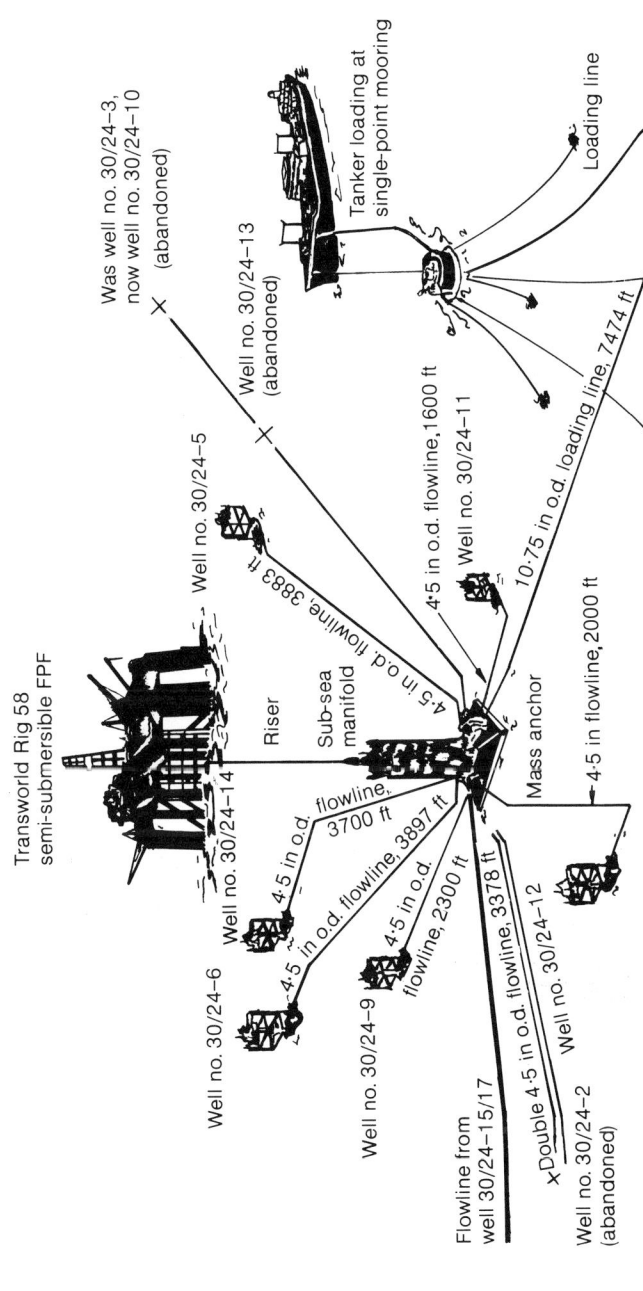

Fig. 9. Argyll Field installation

Fig. 10. Transworld 58 rig

Fig. 11. SBM mooring buoy and tanker

Separator fluids are drained in order to decrease barge deck
load and fluid surface which would cause a mobile 'slopping'
load as the barge pitches and rolls during bad weather.

66. Occasionally, if in the judgement of the Field Super-
visors, the weather is not going to worsen past this level,
then in consultation with the tanker Captain, it may be
decided to remain on the mooring. In this case instead of
leaving at once the tanker will disconnect the loading hose
and hang onto the hawser. This allows the mooring to be
dropped immediately that decision is required. If this
judgement has been correct then as the mooring tension drops
when the weather abates, all that is required is for the
hose to be reconnected and production recommenced.

67. Assuming that the tanker released it's mooring and the
weather continued to build then the next phase of the operation is to monitor the barge heave and anchor chain tensions. These give a good indication of the effect of the
weather. An anchor chain tension of 150 Kips (lbs x 1,000)
equivalent to 67 long tons, or a heave of 8 ft. is the maximum parameter at which production riser disconnection and
pulling should commence. The riser goosenecks are disconnected and the pipes pulled and stood back in the barge derrick. The 10 in. riser is pulled similarly until the universal joint is able to be hung off from the rotary. When
the 10 in. riser is disconnected from the manifold, the
barge is able to be ballasted down to make it more stable.

68. If the weather continues to worsen then the two 42 line
control pods are disconnected from the subsea manifold after
closing the down-hole-safety-valves in each well and pulled
to surface. This is conditional on the deck conditions
being safe for working and is not normally done, as the
flexible hoses are able to tolerate severe rig movement
without damage.

69. Restarting production follows the reverse of the above
events. It is usual, with the aid of the manifold subsea
television system to rerun and latch the riser system
before the tanker is able to remoor.

70. Prior to mooring the tanker a visual check is made of
the floating loading hose and hawser to ensure that they are
not entangled, as often happens. With the hawser and hose
'streaming' the tanker is able to moor in up to 15 ft. seas
with 20-25 kt. winds.

71. All subsea flowlines and risers are pressurised with
water before production starts. This serves two purposes,
it establishes the integrity of the subsea pipelines, and
enables the Christmas tree valves to be operated with a
minimal differential pressure, thus reducing wear and tear.

ARGYLL OPERATIONAL DATA
72. The following table (Table 1) and histogram (Table 2) are self explanatory and illustrate date which is typically of interest to those considering the use of a floating production and/or tanker loading facility.
73. The period 1978 to 1981 is taken as being representative. Production and downtime percentages are based on periods of plant utilisation only, they do not take into account individual well downtime.

Table 1. Argyll production summary, 1978-81

	1978	1979	1980	1981
Production Time %	60.83	66.56	67.17	50.97
Principal Causes of Downtime %				
A - TW58 Major Repairs	14.83	0	11.81	21.19
B - SBM Major Repairs	11.80	9.31	0	0
C - Waiting-on-weather	8.66	11.37	11.55	7.82
D - SBM Faults	2.64	11.54	8.92	19.74
E - Production Riser Faults	1.01	0.55	0.36	0.01
F - Production Plant Faults	0.20	0.26	0.06	0.07
G - Control System Faults	0.03	0.38	0.10	0
H - TW58 Faults	0	0.03	0.03	0.20
TOTAL DOWNTIME %	39.17	33.44	32.83	49.03
Occasions Production Riser Pulled	6	3	5	4
Occasions Production Shutdown (By Tanker leaving Mooring because of bad weather)	19	16	22	21

REFERENCES
1. The Argyll Field is operated by Hamilton Brothers Oil and Gas Limited on behalf of the following consortium of Companies.

   Hamilton Oil G.B. P.L.C.
   Rio Tinto Zinc Oil and Gas Limited
   Texaco North Sea U.K. Limited
   Blackfriars Oil Company Limited
   Hamilton Brothers (U.K.) Petroleum Corporation
   The Trans-European Company Limited

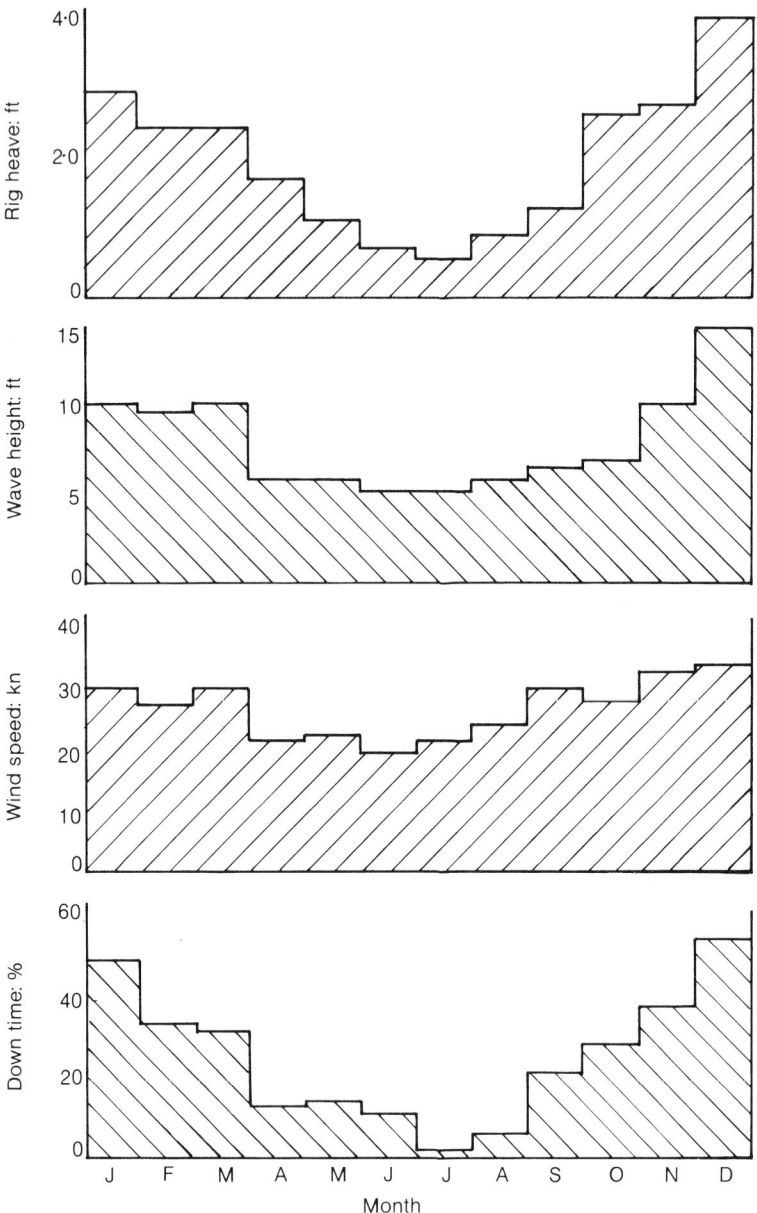

Table 2. Histograms of average maximum data, 1978-81

4   Offshore catenary moorings

P. A. JORDAN, BSc(Eng), FIMechE, MICE, Seaforth Maritime Ltd
and R. W. BREWERTON, BSc(Eng), MICE, Foster Wheeler
Petroleum Development Ltd

Much of the Authors' experience on which this paper is based
was gained whilst they were employed by Shell UK Exploration
& Production.   The opinions expressed are, however, the
Author's own and do not necessarily reflect those of Shell,
Seaforth Maritime Limited or Foster Wheeler Petroleum
Development Limited.   The Authors wish to thank Aker
Engineering for their assistance in providing data on their
Aker H3 and H3-2 Vessels.

SYNOPSIS. Design and construction of offshore Catenary
Moorings for production facilities are discussed.   Some
practical considerations in respect of installation of such
moorings are indicated.   Safe design of moorings can be
carried out in a relatively simple manner, and a worked
example is included.

INTRODUCTION

1.   It is likely that mooring equipment was invented soon
after the invention of the first boat.   Mooring design and
practice has developed empirically over the years without
any dramatic change in equipment or procedures.   Improvements that have been made are due largely to the development
of new materials of construction and improved handling
equipment so that heavier wires, chains and anchors could be
handled.   In general, the moorings on ships are subject to
intermittent, short-term use so that inspection for wear,
and replacements of components can be carried out easily
using the equipment on board the vessel.   The mooring of
large permanent offshore structures in exposed locations
required a different type of service from a mooring.   The
moorings are permanently in use and so consideration of
fatigue, brittle fracture and wear have to be considered.

The forces generated are large, and heavy moorings are
required; at the same time the permanent nature of the
moorings often make it undesirable to install winches for
these heavy moorings on board the floating unit and other
solutions have to be considered. This paper considers the
various aspects of this extrapolation of mooring technology
into the offshore structures field.

Behaviour

2. Before considering the behaviour of Catenary Moorings,
it is as well to think for a moment of the reasons for their
use. Catenary Moorings are used for floating structures
which are in themselves used where they are considered to be
more economical than fixed structures. Catenary Moorings
are redundant and flexible and provide a progressive, elastic
response to environmental forces. The floating catenary
moored unit thus absorbs and dissipates the energy from the
environment by means of a damped elastic spring system.
This means, of course, that the unit can and, indeed, must
move within the restraints of the mooring system. Since
most floating structures for oil-field use have to be
connected to the sea bed, either by a marine riser or with
hoses, the design of those flexible tubular components
provides a limitation to the distance which a floating unit
can be allowed to move. One of the principle tasks of the
designer of Catenary Moorings is, therefore, to ensure that
the system characteristics are such that the movement of the
floating unit under extreme environmental conditions will be
suitably small.

Design

3. The principle aim inthe design of a Catenary Mooring
System is to ensure that the floating unit remains, within
acceptable limits, in position whilst subject to the
environmental forces which occur at its location. These
forces are induced by waves, wind and current. Wave forces
are very important. Two forms of wave forces occur, first
order wave forces which induce a movement in the unit at the
same frequency as the waves themselves and the second order
force which would be a constant force in a regular sea but
in real, irregular seas, result in a slowly varying drift
force. Wind force is also a significant source of loading
on offshore structures. Wind forces also vary in an
irregular cyclic manner and there is some evidence that there
are second order slowly varying wind forces also. In the
case of semi-submersibles wind forces are relatively more

more important than wave drift forces.   Considerable work
has been done in recent years to develop analytical methods
of predicting the extreme forces on floating units.   Most
existing moorings have been designed using the so-called
Quasi-Static method in which combinations of wind, wave and
current forces are calculated and applied as static forces
to the unit, ignoring the low-frequency drift forces.   The
rather conservative assumption that the maxima of these three
forces coincide both in time and direction is made.   The
additional loads due to LF drift motion can usually be ignored.
There is a good deal of work in hand at present to develop
computer programmes which will give a dynamic, time-domain,
simulation of forces and motions on a floating vessel.   These
methods attempt to make an assessment of the slowly varying
drift forces on the unit but as with any forced-damped spring
mass system, the effects of the slowly varying drift forces
will only have a severe effect on the excursion amplitude if
the exciting frequency is near to the natural frequency of
the system.   In this case, the displacement amplitude will
become critically dependant on the viscous damping of the
system.   In Appendix I the Quasi-Static method is applied
to a design for a mooring of an Aker H3-2 in deep water.
The low frequency motions though large in themselves are not
sufficient, when combined with other load effects, to
generate higher design forces than given by the conventional
Quasi-Static method.

4.   The calculation of forces in catenaries and the way these
forces vary as the geometery of the catenary changes, is
laborious rather than difficult.   Data is available for
calculating catenaries in tabular form and they can be
readily calculated using a programmable hand-held electronic
calculator.   The calculation of the forces in a mooring
system using a tabular method is included in Appendix I to
this paper.   It has been found that when a floating offshore
unit is installed, there are often minor adjustments to be
made to position it correctly over the sub-sea manifold.
In these circumstances, it is laborious to use an iterative
process of altering chain lengths, viewing the result and
re-calculating the next alternative.   To simplify this
matter a nomogram has been produced, an example of which is
shown as Appendix II.   This nomogram enables one to quickly
assess the number of links which a chain has to be moved to
produce a required effect on the chain tension and the
position of the unit.

SESSION 2

MATERIALS
Chains

5. Scrutinization of the calculations of catenaries shows
that chains with their high weight per unit length often have
a higher energy absorption capacity that a wire rope. Whilst
chain has this very desirable property of giving a good energy
absorbing catenary, it unfortunately suffers from that well
known failing of being only as strong as its weakest link.
In order for chains to be economically produced, mass
production methods have to be used. Many chain manufacturers
use semi-automatic methods in which the quality of the
finished product is dependant on the skill and diligence of
the operators. In many cases, the quality depends on the
subjective opinion of the operator and will thus be quite
variable. In order to overcome this difficulty, anchor
chains are proof tested as a final stage in manufacturing.
The proof test is a tensile loading of a 90 foot length of
chain to a load well in excess of that to which it will
normally work.

6. A large permanent installation in the Northern North Sea
has a composite mooring consisting of 4" chains and 89mm
diameter wire ropes connecting these chains to the anchors.
The chains are of U2 quality and were manufactured to the
requirements of Lloyds Classification. At that time, a
number of problems were being experienced with chains used
for the mooring of drilling rigs breaking in service. Prior
to installation 100% of the flash-butt welds were inspected
by an ultra-sonic process. At the same time, calculations
were carried out which showed that fatigue was not a likely
mode of failure.

7. Consideration was given to the possibility of failure
by brittle fracture and COD tests were carried out by the
Welding Institute on specimen links removed from the chain
to determine the fracture toughness from which an acceptable
defect size could be computed. The ultrasonic inspection
revealed that in a total length of some 1.5 km of chain,
7 links were found to have defects in excess of the acceptable
size. These links were removed and placed by Kenter Shackles.
Recent work by DNV (Reference 1) has independantly confirmed
this finding that brittle fracture is a likely mode of failure
of common links and fatigue an unlikely mode of failure. In
the same Paper, it is noted that Kenter Shackles, by virtue
of the sharp edged grooves machined in them, are liable to
fatigue failures, a fact that has been independantly verified
by full-scale testing by Shell. It can be concluded,

therefore, that in order to give reliable service for an
offshore mooring, some modifications to conventional chain
manufacturing and inspection procedures are desirable.   Some
chain manufacturers are, apparently, moving towards an
automated process which should give more consistent quality
of welds.   Where this is not possible, more stringent manual
quality control could advantageously be introduced.   Every
part of the chain link manufacturing process should be carefully controlled but, in particular, the formation of the
main flush-butt weld should be carefully watched.   The use
of shackles for joining links of chain is necessary because
longer lengths of heavy chain than those currently produced
would be difficult to handle, especially during proof testing.
Kenter Shackles are the favourite type for this purpose since
they can be joined direct to common links and can in general
be used to pass through the same fairleads, gypsies, etc
as the conventional common link.   Nevertheless, they do
constitute a risk point in service, not only from the point
of view of fatigue, but also because it is not unknown for
them to fall apart in service especially when the chain is
subjected to violent motions.   This is more of a problem
with SBM's than with the larger moored structures here
considered.   There would seem to be some scope for reviewing
the design of Kenter Shackles to see whether machining could
be designed with lower stress concentration factors.

8.   One point which should be considered in the use of chains
is the design of fairleads.   There has been some recent work
done on establishing the effects of tensioning a chain over
a curved surface (Reference 2) in which the conclusion is
reached for ungrooved fairleads the diameter of the surface
over which the chain is tensioned should not be less than
seven times the diameter of the chain.

Wire Ropes
9.   Wire ropes are an alternative to chains.   A wire rope
is much lighter than a chain of the same tensile strength
and thus has less energy absorption capacity by catenary
action.   Wire rope is, however, much more elastic than chain
and this often means that its response is as good as chain of
the same strength.   It is quite possible to make catenaries
which are a combination of chain and wire rope, with the
chain adjacent to the moored unit and the wire lying on the
sea bed and connected to the anchor.   This arrangement
combines the catenary qualities of chain and elastic qualities
of wire and minimises dynamic loads.   In these circumstances,
under moderate weather conditions, the whole length of wire

rope lies on the sea bed and the mooring behaves as if it
consisted entirely of chain.

10. Wire ropes are manufactured in long continuous lengths
by twisting together large numbers of drawn steel wires.
The wire drawing process in itself ensures that there are
very few defects in the wire. The large number of wires in
a rope ensure that failures of a number of individual wires
can be tolerated without any significant degradation in the
performance of the rope as a whole.

11. The result of galvanising the individual wires and
packing the wire rope with wax is that deterioration of the
wire rope eventually starts on the outside where it is easily
seen during inspections. Wire ropes can suffer from bending
fatigue failures where they pass over fairleads. Fatigue
causes its initial failures in the outer layer of strands
where it can be visually detected. After several years in
service a $3\frac{1}{2}$" diameter mooring wire of an offshore loading
unit was retrieved after its accidental severance. Whilst
some reduction in thickness of the zinc coating had occurred
on the outside of the rope, the interior was found to be quite
unaffected by its period in service and the strength of the
rope was unimpaired. Rope manufacturers have already done
considerable work in developing techniques for design,
fabrication and inspection of large sized ropes for future
use with TLP's and Guyed Towers as well as traditional
applications (Reference 3 and 4). This development of wire
rope technology could, of course, be equally applied to
Catenary Moorings.

CATENARY TERMINATIONS AND INSTALLATION

12. In general, Catenary Moorings for moored offshore
production systems use traditional materials although
frequently in large sizes and manufactured to higher
standards of quality assurance. In the case of the ship or
the semi-submersible vessel which is a mobile unit, the
mooring wires or chains are normally handled by means of
winches. These winches allow the mooring to be paid out
rapidly during anchor laying, to store the chain or wire
when it is not in use and to adjust its length and tension
as necessary in service. These winches occupy a large amount
of space and are a considerable weight. In any kind of
floating craft, space and weight are always at a premium
and in the case of a permanently moored production facility,
there is neither real requirement for rapid deployment of

the wire or chain nor any necessity to store it all on board
the unit. It will, however, always be necessary to have the
ability to adjust the length and tension of catenaries. This
can be carried out by linear winches which are relatively small
in size and light in weight. These winches, whilst supplying
the necessary force to the cable, are of low power and hence
move the cable slowly. This type of system has considerable
economic advantages but makes the job of installation of the
mooring system very much more difficult than the conventional
laying of anchors by an offshore work vessel. A further
disadvantage is that since the rate of movement of such a
winch is very slow, it is, generally, unsuitable for embedment
of ship's type anchors, so pile or gravity type anchors will
have to be pre-installed. Use of such an anchor, which can
be installed with a high degree of positional accuracy is
also advantageous in ensuring optimum performance from the
mooring system as a whole and is certainly a necessity if
ever a mooring system in which each leg consists of a
combination of different types of sizes of materials is
contemplated.

13. A floating production system will normally be completed
inshore and towed to its working location by several tugs.
It is a well established practice for such tows to rearrange
the position of the tugs in a star pattern when the unit
arrives at location to enable it to be accurately positioned.
If fixed, pre-installed, anchors are used in combination with
low-power linear winches on the unit, it will be necessary
to pre-lay the moorings prior to the arrival of the tow.
The catenaries will, of course, be much longer than the
direct horizontal distance from the anchor to the centre
point of the production unit. It will, therefore, be
necessary to lay the moorings on the sea bed from the anchor
in towards the centre of the pattern but to either double-
back or alternatively (if chain is used) to leave the catenary
material in a heap on the sea bed. Naturally, the free end
will have to be anchored and buoyed off to permit easy handling.
The laying of too many lines in this manner adds considerably
to the danger of damage and entanglement of the lines, and
these considerations will normally mean that only three, or
four lines can be laid in this way. If there are pre-existing
sea bed installations such as a manifold or pipelines, problems
are even more complicated and very careful positional control
of vessels during laying operations is required. When the
production unit has been moved into position by its flotilla
of tugs, the anchoring vessels will have to approach the unit
in between the tugs in order to connect up the catenaries.

Several different ways of carrying out this aspect of the
work can be conceived, depending upon the type of connection
which has to be made. The pulling of the two ends of cable
onto the deck of an anchor handling vessel in order to make
a connection with the vessel close alongside the unit and
in between the tow lines of the tugs, means that this has to
be a fine weather operation. In the event that the timing
of construction of a unit is such that the installation has
to be carried out in potentially bad weather (eg between
September and March in the Northern North Sea) considerable
care will be required to pre-plan the marine operation and
to ensure that adequate craft, equipment and personnel are
available. If the space constraints at the centre of the
mooring pattern are such that only a fraction of the total
mooring can be pre-laid, then the laying of the other
mooring legs after the preliminary connections must be
arranged to take place fairly quickly, especially if severe
weather is foreseen. The matter can be expedited by pre-
laying parts of all the moorings so that only relatively
short lengths of chains or wire have to be added to those
to be connected after the tugs have departed. The
proliferation of buoys marking the ends of these wires may,
however, make the manoeuvring of the unit into position more
difficult. Once the moorings are connected, a considerable
amount of slack will have to be taken in, and if this has to
be done with the linear winches, it will be a rather long
laborious operation. A temporary addition of an additional
winch enabling the catenaries to be pulled through and
partially pre-tensioned could save a considerable amount of
time. Experience has shown that when moorings are laid in
this way it is very difficult to straighten them out due to
the friction of the chain or wire laid loosely on the sea
bed. A necessary part of any installation procedure should,
therefore, be to tension up to the highest possible load
each of the mooring lines. This is best done by tensioning
each leg in conjunction with the diametrically opposite one.

14. It should be understood from the outset that if it is
 intended to take advantage of the permanent nature of the
 mooring in order to save space and weight on the unit by
deleting conventional winches, then the installation of the
mooring will become an elaborate marine operation which
requires considerable pre-planning. It is quite likely that
early identification of the vessels to be used will be
required, since modifications may have to be made and special
equipment designed, manufactured, installed and tested on

board. The economic effects of committing early for such
a vessel and the costs of the design work and the special
equipment involved must, of course, be weighed against the
economic advantages of the economies possible on the
production unit.

BROKEN MOORING LINE

15. A typical catenary mooring system designed for 100 year
storm conditions can also withstand at least the 10 year
storm from any direction with one line broken.   If a line
breaks during the 100 year storm the adjacent lines will be
subject to a shock load as the vessel shifts to a new
median position.   It is important, therefore, to pay
attention to details to ensure that this excursion and load
can be absorbed if necessary without premature failure of
fair leads and risers.

In the event that a line breaks an adequate temporary
replacement anchor line can be installed in a few days
providing suitable shackles are available.   Alternatively,
as a temporary measure the other lines can be adjusted to
reposition the unit and restrict its excursions.

APPENDIX 1

Example Calculation of Maximum Loads in an 8 point Mooring
For a Semisubmersible Drilling Vessel in 155 metres of Water

Environmental conditions assumed are typical of one year
return storm conditions in the Northern North Sea.

Vessel Type

16.   Twin pontoon semisubmersible with circular columns.
Displacement 22000 tonnes at survival draft.   It will be
assumed that the mooring is permanent and that no attempt
is made to reduce loads by use of propulsion machinery or
by adjusting chain tensions during the storm.

Mooring Pattern

17.   8 Point mooring.   Each line comprising 850 metres of
3" chain and 800 metres of $3\frac{1}{2}$" diameter wire.   Wire at
anchor end.   Pretension at fairlead 50 tonnes.

Mooring Chains

18.   Type   3 inch ORQ
       Wt/metre submerged: 117 kg/m
       Break load: 472 tonnes
       Proof load: 313 tonnes
       Water depth below fairlead: 150 metres
       Length of chain from fairlead: 850 metres
       Effective elastic modulus Kg/cm^2 7800 Kg/mm^2
                                        (on gross CSA 14.1"^{2})
       Elastic strain at breakload   0.663%
       Coefficient of friction of chain   0.75
       on seabed

Wire Rope

19.   Length to anchor   800 metres
       Size/construction $3\frac{1}{2}$", 6x41
                                                          construction
       Breaking Strength   498 tonnes
       Elastic Modulus   7000 Kg/mm^2
                                                          on nominal CSA
       Elasticity of chain wire   32.7 T/metre
       combination   extension

Chain Catenary Calculations   (See figure 1)

20.

TABLE 1   Geometry & Forces – Applicable at all Water Depths

Θ degrees	$\dfrac{L}{d}$	$\dfrac{h}{d}$	$\dfrac{L-h}{d}$	$\dfrac{F}{qd}$	$\dfrac{H}{qd}$
90	1.C0	0	1.0000	1.00	0.00
60	1.73	1.32	0.4151	2.00	1.00
55	1.92	1.55	0.3684	2.35	1.35
50	2.14	1.82	0.3258	2.80	1.80
45	2.41	2.13	0.2864	3.41	2.41
40	2.75	2.50	0.2495	4.27	3.27
35	3.17	2.96	0.2146	5.53	4.53
30	3.73	3.55	0.1813	7.46	6.46
25	4.51	4.36	0.1493	10.67	9.67
20	5.67	5.55	0.1183	16.58	15.58
15	7.60	7.51	0.0889	29.35	28.35
12.5	9.13	9.06	0.0732	42.20	41.20
10	11.43	11.37	0.0584	65.80	64.80
07.5	15.26	15.21	0.0437	117.00	116.00
05	22.90	22.87	0.0291	263.00	262.00

SINGLE LINE CHARACTERISTICS

21.

d=150m   q=0.117 T/m   qd=17.55 T
E=150 - (L-h)   EE=H ÷ 32.7 (m)
(catenary effect)   (elastic effect)
ET = E + EE = Total Excursion

TABLE 2

Θ	L-h (m)	E (m)	H (T)	F (T)	EE (m)	ET (m)	h (m)	L (m)
90	150	0	0	17.55	0	0	0	150
60	62.26	87.74	17.55	35.10	.54	88.28	198	260
55	55.26	94.74	23.70	41.24	.73	95.47	232	288
50	48.87	101.13	31.60	49.10	.95	102.08	273	321
45	42.96	107.04	42.29	59.84	1.31	108.35	320	362
40	37.42	112.58	57.39	74.94	1.76	114.34	375	413
35	32.19	117.81	79.50	97.05	2.42	120.23	444	476
30	27.20	122.80	113.37	130.92	3.46	126.26	532	560
25	22.40	127.60	169.71	187.26	5.19	132.79	654	676
20	17.745	132.25	273.43	290.98	8.36	140.61	832	850

22.

TABLE 3 Design Environmental Conditions

Wind:
1 hour Mean (25 M above MSL)		34.0 M/sec
10 min mean		1.03 x 34 m/sec
1 min mean		1.14 x 34 m/sec
30 sec mean		1.19 x 34m/sec

Wave:
H_s	12.1 m
H max	22.4 m
T_z	11.1 - 12.5 sec
T_s	13.3 sec
$T_{ass.}$ assoc with H_{max}	15.3 (13.5 - 17.1)sec

Current:
Surface	1.10 m/sec
at 15 m depth	0.84 m/sec

TABLE 4 Mean Environmental Forces (tonnes)

Weather Direction	Bow	Quarter	Beam
Mean hourly wind	88	88	88
1 min mean wind	114	114	114
Wave drift	27	27	27
Current	46	65	95
Total (mean wind)	161	180	210
(1 min gust)	187	206	236

23. High Frequency Motion

This is caused by first order wave effects. A surge and
sway response spectrum for the vessel is given in figure 3.

We only require the 3 hr maximum value. The shape of the
response spectra is such that these values can be obtained
directly as follows: H max = 22.4 m T_{ass} = 15.3 sec

from figure 3 find R associated with T_{ass}: $2A_H = R H_{max}$
where A_H is the 3 hr maximum single amplitude of H F motion.

surge motion R = 0.75	$A_H max$ = 8.4 m
sway motion R = 0.69	$A_H max$ = 7.7 m
quartering motion R = $\sqrt{0.53^2 + 0.49^2}$	$A_H max$ = 8.2 m

24. TABLE 5   Excursion and Mooring Forces
                 (determined from Figure 4)

Weather Direction	Mean Force	Excursions: Mean E_s	Dynamic E_d	Total E_t	Chain Max Force F(Tonnes)
i) Intact Mooring					
Bow	187	24.2	8.4	32.6	210
Quarter	206	26.0	8.2	34.2	230
Beam	236	28.0	7.7	35.7	250
ii) Damaged Mooring (Windward line broken)					
Beam	236	44.8	7.7	52.5	265

25. Load Effects Not Evaluated

a) Vessel Heave
This does tend to increase chain loads but not by a significant
margin since heave motion is out of phase with surge motion
and the chain angles at the vessel are low under maximum load
conditions.   In this design, considerations of heave would
have increased design loads by less than 10 tonnes.

b) Drag of Chains Through the Water
This can have a considerable effect in deep water situations
and will reduce the catenary effect.   Large high frequency
changes in the catenary geometry will be resisted due to the
implied high lateral chain velocities.   (6 to 10 m/sec in
this case)   This results in a supplementary cyclic load in
the chain and a higher effective mooring stiffness for high
frequency motion effects.   The extra elasticity provided by
the wire rope section will help to minimise this effect.

c) Unsymmetrical Windage

If the centre of pressure is away from the centre of gravity
of the mooring pattern, the vessel will yaw and this will
result in uneven loading of chains.   In this case, the vessel
is symmetrical and this effect can be ignored.

(ERRATUM: The Authors wish to draw attention to Fig. 4 of
their Paper and point out that although the curves and mean
excursions scale correctly, values of dynamic excursion 'Ed'
and maximum loads are too large; the values in the text are,
however, correct.)

25. Load Effects Not Evaluated

d) Inaccurate Adjustment of Chains Pretension

The effect of this is normally negligible in a well installed mooring but good in-service monitoring procedures are also necessary.

e) Low Frequency Drift Forces

Space limitations preclude a detailed presentation of these effects. In this particular example consideration of L.F. drift motions did not lead to higher design loads.

REFERENCES

1. Berg A and Taraldsen A, DNV "Long term mooring and anchoring of large structure and drilling units. (Reliability and Safety of Anchor Chain System) OTC 3813.

2. Flory, J F and Woehleke, Exxon "Strength of Chain Tensioned over a curved surface" OTC 3855.

3. Ronson, K T British Ropes Limited, "Ropes for deep Water Mooring" OTC 3850.

4. Sharp, D M "Rope in the Marine Environment" I Mech E Proc 1976 Vol 190 5/76

5. Borresen, J and Olsen, E "Ross Rig Instrumentation & Data processing, Norwegian Meterological Institute Environmental Data Centre".

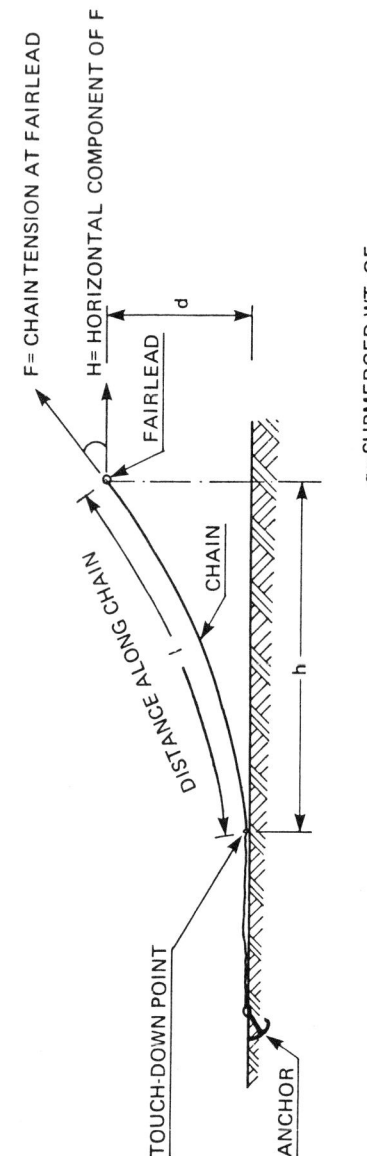

Fig. 1. Chain catenary calculations

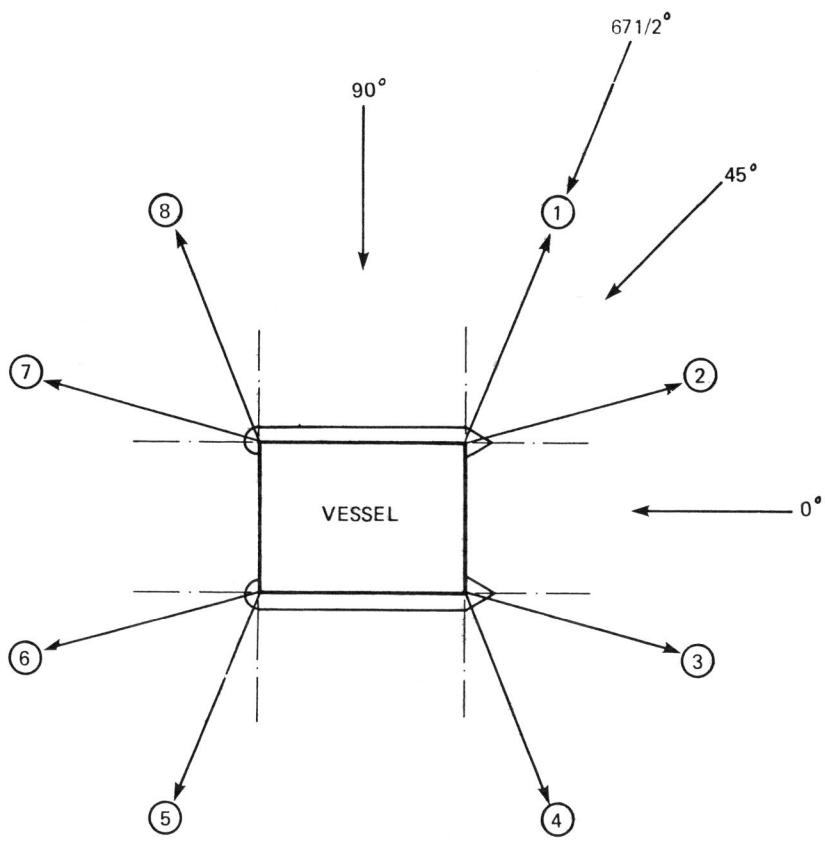

Fig. 2. Mooring pattern

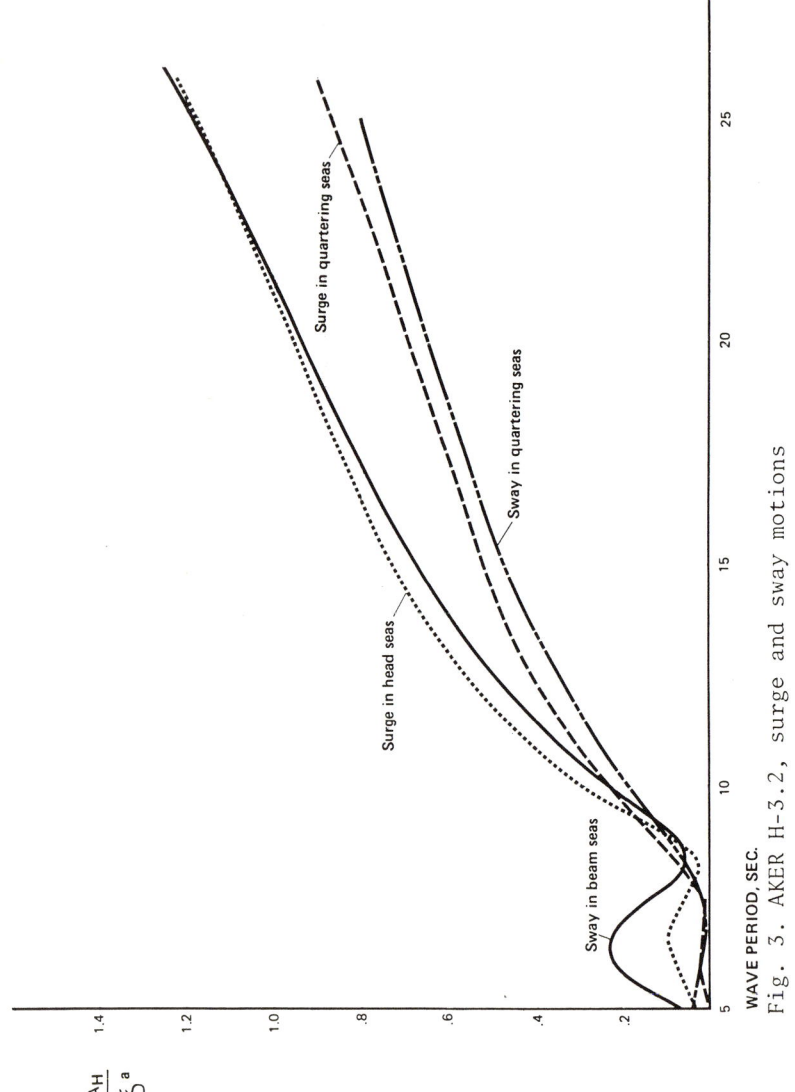

WAVE PERIOD, SEC.

Fig. 3. AKER H-3.2, surge and sway motions

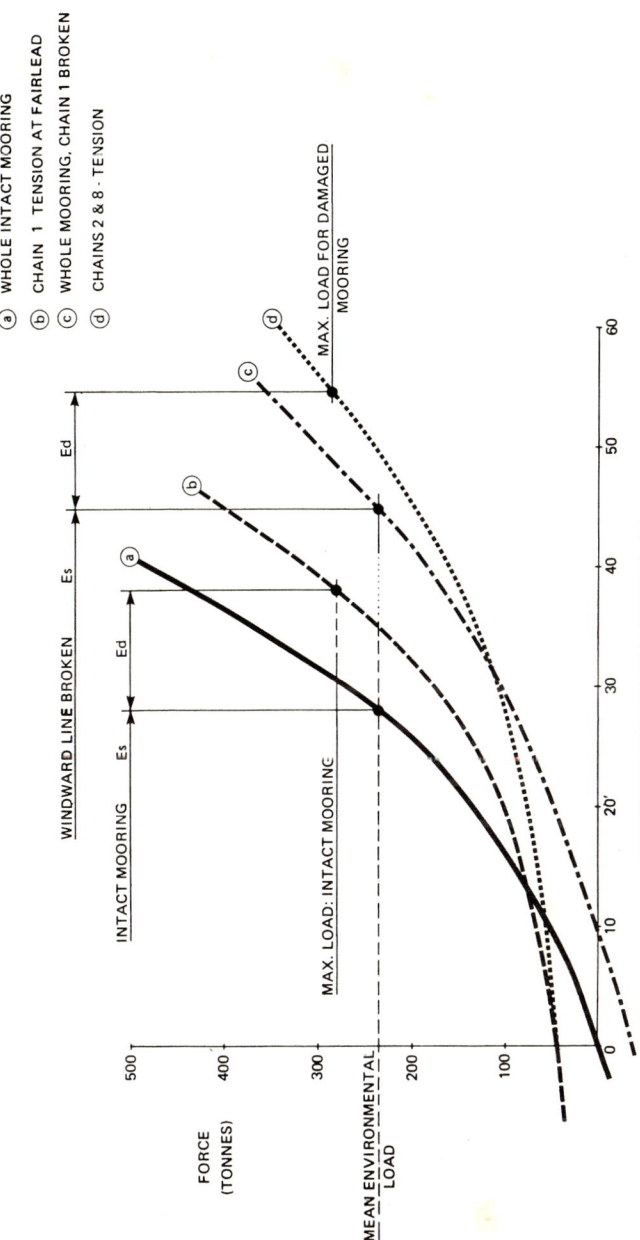

WEATHER DIRECTION 221/2 DEG. OFF BEAM
1 YEAR STORM
LOAD COMBINATION 1
LOAD EXCURSION CURVES:

(a) WHOLE INTACT MOORING

(b) CHAIN 1 TENSION AT FAIRLEAD

(c) WHOLE MOORING, CHAIN 1 BROKEN

(d) CHAINS 2 & 8 - TENSION

Fig. 4. Forces and excursions

113

APPENDIX 2

Nomograph to assist chain tensioning for loading buoy

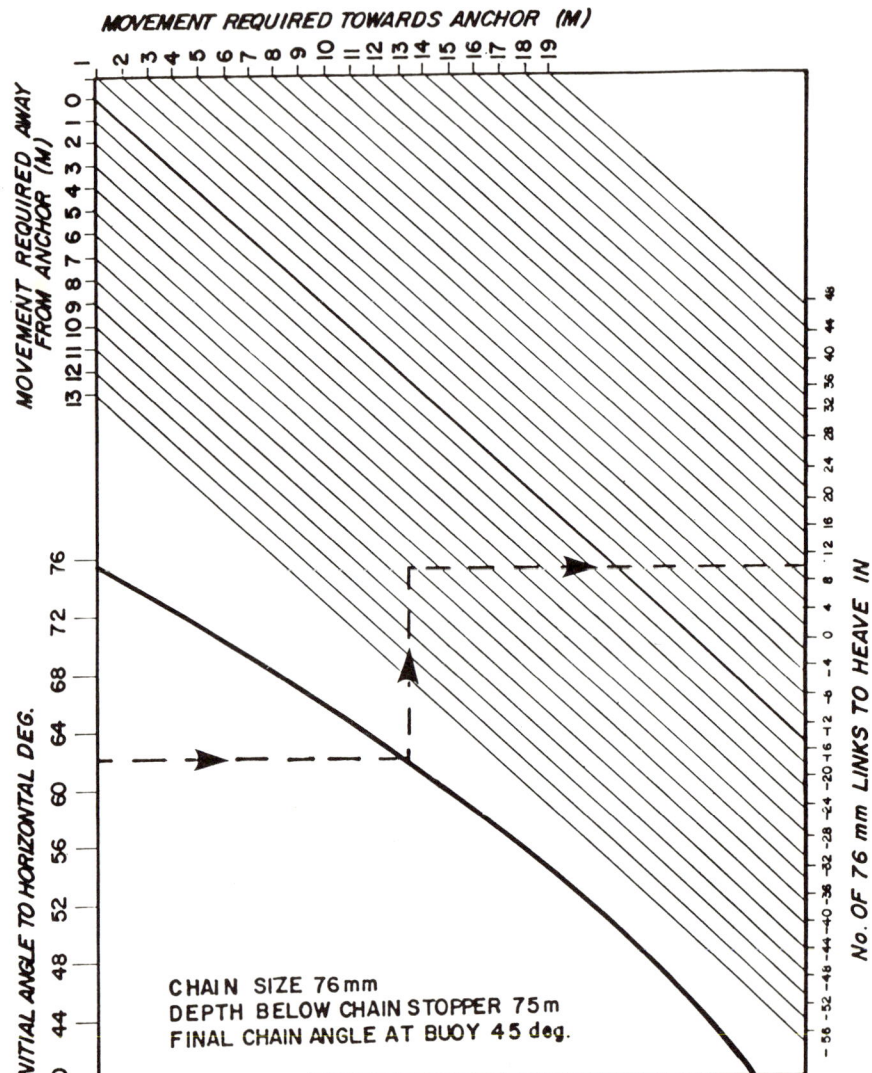

MOVEMENT REQUIRED TOWARDS ANCHOR (M)

MOVEMENT REQUIRED AWAY FROM ANCHOR (M)

No. OF 76 mm LINKS TO HEAVE IN

INITIAL ANGLE TO HORIZONTAL DEG.

CHAIN SIZE 76mm
DEPTH BELOW CHAIN STOPPER 75m
FINAL CHAIN ANGLE AT BUOY 45 deg.

Discussion on Papers 3 and 4

MR G. DE LAVAL, *Bulten-Kanthal AB, Ramnäs Division, Sweden*

1. The Authors of Paper 4 quite correctly quote from reference 1 (OTC 3813, 1980) that brittle fracture has been the most common reason for chain breaks in the North Sea. I wish to emphasize how primitive the brittle fractures are and how relatively easy they are to avoid by discrimination at the time of purchase. In order to give an idea of what is meant by reliability of chain, I will give details of the Ramnäs quality system.

2. The most important criterion for quality manufacturing is to have a production technique that is simple and which makes it as easy as possible to do the right thing. The Ramnäs linear method of chain-making is in many respects an excellent foundation for high-quality performance. The chain is forged continuously and the three-link samples are easy to relate to the corresponding chain. It is difficult to make errors and I think that this is of major importance.

3. The continuous procedure has many other advantages: no repair links between shots; lower costs due to the absence of shackles; and greater safety for the same reason. Incidentally, can the Authors of either Paper explain why ship-owners do not order continuous chain?

4. In manufacture, the linear layout makes it easy to put the machines in such positions that they are readily accessible for setting and for repair, and particularly for automation.

5. The Ramnäs linear method of chain-making is practised throughout the whole manufacturing process, e.g. Ramnäs has pioneered continuous testing which makes it very difficult to do anything wrong as regards relation of samples to

Table 1. Contents of Ramnäs quality assurance manual for
anchor chain and accessories

---

1. Manufacturing facilities, process description
2. Organization
3. Distribution of responsibilities
    (a) decisions on quality
    (b) specifications for: purchase and manufacture
    (c)   checking, testing and inspection
    (d) education and training of personnel
    (e) internal audits
    (f) statistics
    (g) documentation: issuing, monitoring, updating, re-
        calling
4. Specification for: bar material, studs and accessories
5.   reception of material according to 4.
                       above
6.   identification through manufacture
7.   bar-cutter setting
8.   checking blanks
9.   setting of heater
10.   checking heater
11.   setting of bending machine
12.   checking of hot link shape
13.   setting of welding machine
14.   checking welding result
15.   trimming machine setting
16.   checking trimming
17.   setting of stud press
18.   checking hot link shape
19.   heat treatment
20.   checking heat-treatment result
21.   mechanical testing
22.   load testing
23.   visual inspection
24.   ultrasonic inspection
25.   magnetic particle inspection
26.   surface coating
27.   maintenance of manufacturing machinery
28.   maintenance of checking and measuring
                       instruments and test equipment
29.   handling of non-confirming material
30.   corrective actions

---

the corresponding lengths of chain. With continuous testing
it is also possible to obtain a better pitch control than
with the batch-type testing method.

6. The following are details of Ramnäs production flow which
all contribute to consistent high quality:

(a) precision cutting with cold circular saw immediately
    before forging
(b) automatic temperature control of heater
(c) regular check of welding parameters by diagram and by
    practical tests
(d) mechanical trimming which makes burr removal independent
    of human strength. If the chainsmith has to chisel the
    burr manually, he will set the welding machine in a
    manner that will produce a small burr; this method has
    ruined many links. A generous flashing length and a
    large upsetting stroke are 'musts' in order to obtain
    good welding quality. Heavy links with good welds have
    very large burrs which are more or less impossible to
    trim manually, on larger sizes of chain. Automatic
    trimming, combined with thorough flashing and upsetting,
    creates the foundation for a sound flash weld
(e) the fully automatic function of all machine operations
    in the Ramnäs factory improves consistency to a high
    degree and is a good safeguard against the human factor.
    Ramnäs utilizes men only to check that the machines have
    not failed. Men are never put in a position to operate
    the machines incorrectly because everything is computer-
    ized and the settings of the machines are well documented
    and controlled.

7. The second major feature contributing to a good product
in our factory is the Ramnäs quality assurance programme.
This is described in the quality assurance manual, the contents of which are shown in Table 1. Within the programme,
I believe that the preventive maintenance of tools and
machines is by far the most important quality factor.

MR D. H. ROBERTSON, *Marine Technology Support Unit, AERE
Harwell*

8. In relation to the failed links in the Transworld 58
incident, unless failure occurs at the flash butt-weld,
there is probably a good chance that the stud will retain a
broken link at one end of the chain after it has parted.
Since all the mooring chain was salvaged after the incident,
can Mr Cockrill say if any of the broken links were recovered
from the six cables which had parted? If so, what failure
mechanism has been attributed to these links?*

SESSION 2

DR D. G. OWEN, *Department of Offshore Engineering, Heriot-Watt University*

9. With the benefit of hindsight, would the Author of Paper 3 have adopted a different course of action with regard to the Transworld 58 incident?*

REAR ADMIRAL A. F. R. WEIR, *G. Maunsell & Partners*

10. With reference to Paper 3, in the light of the question on lessons learned from the Transworld 58 accident, the Author stated that part of the mooring cable had been replaced by 2.75 in cable instead of the 2.0625 in one used previously. Since replacement of only one part of the cable will in no way increase the strength, what was the reason for doing this?*

MR M. ISHERWOOD, *V&O Offshore Ltd*

11. Can the Author of Paper 3 say what environmental conditions were used in designing the Transworld 58 mooring system?*

MR H. CRAWFORD, *National Engineering Laboratory*

12. Can the Author of Paper 3 expand on his experience with hawsers made from polypropylene?*

MR R. S. RAJAN, *Department of Engineering, University of Aberdeen*

13. The Authors of Paper 4 state in paragraph 11 that corrosion/deterioration starts on the outside of a wire rope; this is not always true. In fact, depending on the conditions, deterioration of the rope can take place from the outside or from the core of the rope.

14. Consider, for example, the situation where an anchor cable had been lubricated and put into service; the lubrication is removed in service and sea water penetrates to the core. At a later stage the anchor cable is removed, relubricated and put back into service. The situation now is that the outer surface of the rope is protected by lubrication, etc., but nearer the core the conditions encourage electrolytic action. Hence, in this situation the core corrodes more than the outer wires of the outer strands.

---

* Publishers Note. It is regretted that the Author of Paper 3 did not submit a reply before publication.

15. Another kind of deterioration which can occur on the inside of a wire rope and which is not obvious from the outside is that due to nicking and grooving.

16. With the above in mind and noting the large diameter and length of anchor cables it really is not satisfactory to lay out a length of anchor cable ($\approx$5000 ft) and inspect it visually. Could the Authors indicate what inspection and maintenance procedures they recommend with regard to anchor cables? Can they comment on the use of electromagnetic non-destructive testing of wire rope in relation to anchor cables?

17. In the past, due to the relative cheapness of wire ropes and the relatively short lengths of rope used in service, inspection of wire ropes (if any) has been mostly confined to laying out the wire ropes and visually examining them. It has long been recognized that the size and length of the rope have a direct bearing on the accuracy and effectiveness of visual examination.

18. With the discovery of oil offshore and the advent of tension leg platforms and single-point moorings, catenary moorings and guyed towers, it is no longer viable to visually examine the wire rope due to their size (2 - 4 in dia.) and length (>5000 ft). Therefore it is highly desirable to find or use a technique to measure the level of deterioration of a wire rope which is economical, convenient and accurate.

19. In the mining industry, which is one of the major users of wire rope, various forms of non-destructive testing (NDT) technique for wire rope have been in use for the past 20 years. Of the various NDT methods, electromagnetic NDT is the most effective and widely used method.

20. There are three methods of electromagnetic testing of wire ropes: the AC method, DC method, and one which incorporates Hall-effect sensors.

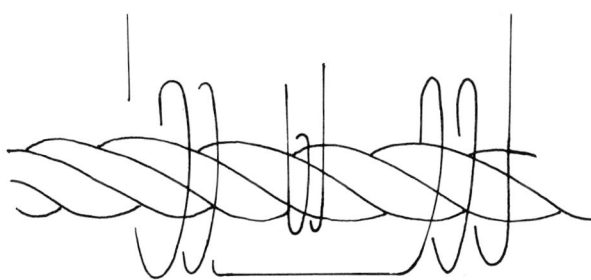

Fig. 1. AC method

21. An instrument based on the AC method resembles a transformer with primary and secondary coils (Fig. 1). The wire rope serves the purpose of the ferro-magnetic core of the transformer. The output voltage of the secondary coil is proportional to the magnetic flux flowing through the section of wire rope, and the magnetic flux flowing through the section of wire rope is proportional to the metallic cross-sectional area of the wire rope. Therefore, variations in the metallic cross-sectional area cause variations in the magnetic flux and voltage, allowing loss in metallic area to be measured by suitably placed sensing coils. Remnant magnetism in the wire rope and the presence of metallic oxides adversely affect the operation of AC instruments. Local faults are not detected.

22. The DC method uses a strong electromagnet or permanent magnet to saturate the measured section of the wire rope (Fig. 2). Local faults distort the magnetic flux passing through the wire rope, and suitable sensing coils detect the leakage flux as the wire rope moves through the instrument. The output signal of the sensing coil depends on the size of the faults and the rope speed. All instruments of the DC type have to be adjusted to compensate for rope speed variations. It is not possible to fully compensate for effects due to change in speed but the major effect is due to broken wires and, since this must be an integer, it is possible to determine this unless many wires are broken at the same position.

23. The NDT instruments incorporating Hall-effect sensors use a strong permanent magnet to apply a static magnetic excitation to the test section of the rope, and the magnetic flux passing through the wire rope is measured using Hall-effect sensors (indicated by arrows in Fig. 3). Simultaneously, leakage flux caused by local faults is also detected. The Hall-effect sensors either replace or supplement coils in conventional testers. Hall-effect sensors can detect and measure static or dynamic fields very accurately. Fine electrical conductors are bonded to all four edges of the sensor (Fig. 4). The other two edges develop a potential difference when the sensor is placed in a magnetic field. Under saturating conditions, the magnetic flux is a linear function of the metallic cross-sectional area of the wire rope; the potential developed by the Hall-effect sensors is in turn a linear function of the flux, thus allowing measurement of any loss of metallic area.

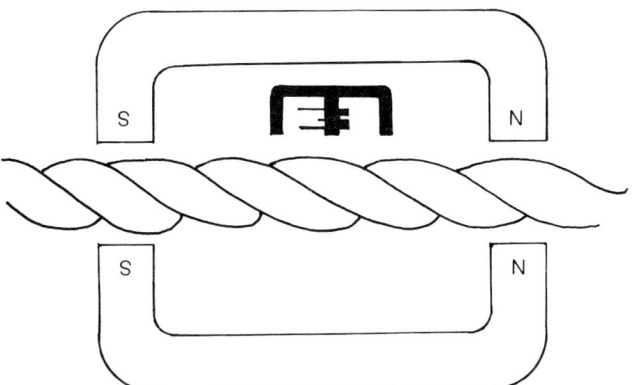

Fig. 2. DC method

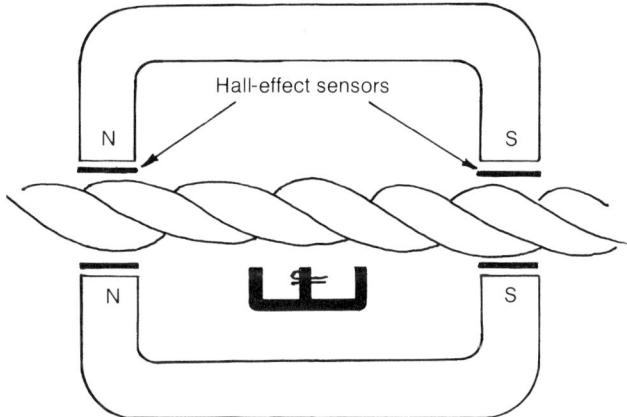

Fig. 3. Location of Hall-effect sensors

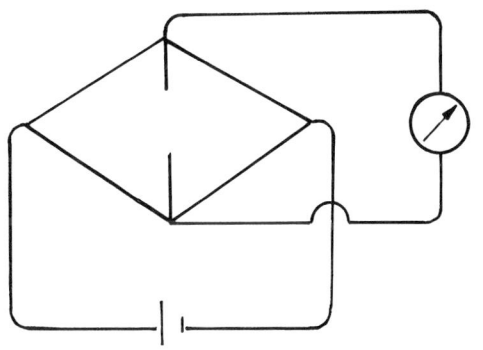

Fig. 4. Hall effect

24. Local faults are detected by a different set of Hall-effect sensors from those which measured the loss in metallic area of the wire rope. In conventional wire rope testers according to Lenz's Laws, current is generated in the sensing coils only when they are moved through a magnetic field; the current generated or the output signal is hence a function of speed and is zero when the rope is at a standstill. The NDT instruments incorporating Hall-effect sensors can measure

(a) the number of wire breaks
(b) loss in metallic area (wear)
(c) gain in metallic area (splice)
(d) corrosion.

25. Only three types of NDT instrument (manufactured by five different companies worldwide) incorporate Hall-effect sensors in their measuring head. Although these NDT instruments have been commercially available for some time, until now no work has been published relating the output from these instruments to results obtained from detailed laboratory examination of the wire ropes.

26. In the Department of Engineering, University of Aberdeen, the following programme of work is intended to relate the records from the NDT instruments on used wire ropes to results obtained from laboratory examination of the same wire ropes. Work has already started on the design and construction of a test rig over which an endless rope with certain man-made defects will be driven. The rig is intended for demonstration purposes and it will also enable researchers to consolidate the necessary expertise in using the instrument which will operate at rope speeds of 0.5-3 m/s.

27. The following programme of work is intended with the co-operation of some offshore operators:

(a) whenever a new wire rope is installed, a 'fingerprint' of the new rope will be obtained using the NDT instrument
(b) periodically, non-destructive testing of the ropes while in service in order to obtain the deterioration characteristic of each rope
(c) whenever a rope or part of a rope is discarded from service (due to normal inspection procedures or results from the NDT instrument) detailed laboratory examination of the worst section of the rope.

MR N. J. HAWKINS, *Hawkins and Tipson Ropemakers Ltd*

28. Discussion has centred on chain and wire ropes which are, of course, most important for platform and rig moorings. However, SPMs are also important, and with most of them a vital link in the mooring system is the hawser between the buoy and the tanker; this is invariably fibre rope - a topic which has received little attention during the Conference.

29. I wish to say that, as fibre rope manufacturers, Hawkins and Tipson have no axe to grind and will make mooring hawsers from any one of a number of fibres. At the moment most of these hawsers are made from nylon, but there have been many instances of sudden failures with nylon hawsers and it has taken a long time for the cause of these failures to be pinned down.

30. We believe that the reason for this is that SPMs are always in a permanently wet environment, whereas rope testing is more usually carried out in the dry environment. Where fibre properties alter little between the wet state and the dry this does not matter, but where there is a large alteration it is, of course, important that the relevant parameters are tested in wet conditions. We, and many other laboratories including NEL, have carried out a great deal of such testing and find that the performance of different fibres with respect to one vitally important parameter, namely, tension fatigue life in wet conditions, is in the following order: polyester; polypropylene; nylon. Accordingly, we are selling polyester and polypropylene hawsers as well as nylon to SPM operators around the world and these operators are finding considerably improved service life.

31. This is important because earlier speakers indicated ways in which the low-frequency tension variation in these mooring hawsers can be minimized, but the wave frequency variation will always be present; this has a cycle time of a few seconds and thus is a major contributor to failure by tension fatigue. It is therefore important to choose the correct fibre, and we believe that the small sacrifice in wet strength involved with the selection of polyester in preference to nylon is more than fully compensated for by the vastly superior wet fatigue life of that fibre; anyhow, if the last 10% of wet strength is regarded as absolutely essential, it is always possible to use polyester in a slightly larger size.

DR E. C. BOWERS, *Hydraulics Research Station*

32. In Paper 4 a calculation of design loads is made using

the 'Design wave' method for wave forces, and low-frequency response is neglected. This is justified on the basis that inclusion of the low-frequency drift forces did not lead to higher design loads when likely combinations of wind, current and waves were considered.

33. The point is also made (see Fig. 4 of the Paper) that if one mooring line breaks then the mooring system as a whole becomes softer and maximum loads in the remaining moorings do not increase. It is worth pointing out, however, that when a mooring line (or lines) does break, the system becomes softer and the resonant frequencies of the structure on its moorings decrease. Reference to Fig. 3 of Paper 1 shows that any low-frequency resonant response will tend to increase since the low-frequency force increases as the slowly varying frequency decreases. Therefore, if mooring systems are to allow for lines breaking without a 'knock-on' effect of more and more lines breaking, then it appears prudent to check that the low-frequency response does not lead to mooring design being exceeded.

MR JORDAN AND MR BREWERTON, *Paper 4*

34. With regard to the contribution from Mr de Laval, we are most interested to read of the manufacturing and quality control procedures adopted by Ramnäs. It would seem that Ramnäs have progressed a long way towards the automated process of manufacture which we advocate in the Paper.

35. In reply to Mr Rajan, in the Paper we were, in general, addressing the situation where a unit is moored, long term, in the same position. Under these circumstances, we believe that our comment that deterioration of the rope starts on the outside is correct; inspection of the rope (which can be carried out by manned or unmanned submersible) should be carried out on a routine basis and the rope replaced if found to be significantly corroded or damaged. We believe that with this type of mooring the lifting of a cable for inspection may result in its damage and, therefore, could do more harm than good. It should be noted that these comments are generally applicable to wire cables lying close to the sea bed in locations like the northern North Sea. Where cables are located in shallower water, conditions of corrosion may be much more aggressive and different procedures may become necessary. In considering cables which are frequently recovered and redeployed, we have a situation much more akin to practice of towing vessels and the like. In these circumstances, it is normal to inspect cables visually from time to time and to lubricate them by spraying with

a suitable lubricant. Inspection of the inner part of the wire is frequently carried out by twisting the wire open using a spike. In experienced hands, this does little damage to the rope.

36. We are currently giving consideration to the required maintenance procedures for anchor cables for a semi-submersible support vessel for use in the North Sea. It is a dynamically positioned vessel and it is anticipated that the mooring wires will be used at rather infrequent intervals. We have seen a device tested which clamps around the cable and allows injection of lubricants under pressure while the cable is being pulled through at a relatively fast rate. Subsequent opening up of the test cable revealed that the lubricant had penetrated all the way to the centre. This method would appear to give a cleaner, more effective and more economical method of lubricating cables than the spraying previously adopted. Visual inspection of the cable can, of course, be carried out at the same time.

37. We believe that cables used in mining, particularly those in hoisting applications, have in many ways more arduous working conditions than anchor cables. They have a very high intermittent usage often in an aggressive atmosphere. Failure of hoisting cables can cause not only loss of life but a very severe effect on the economy of the mine. In these circumstances, on-line inspection of wire ropes seems prudent. It may be that service failures in mining ropes, with their very frequent bending over sheaves, more often start from the interior of the rope. It would seem that development of a suitable electronic wire rope inspection device for use on offshore cables, would be advantageous. In our view, it should have the following characteristics:

(a) be rugged, portable and suitable for use in a marine atmosphere
(b) be programmable to suit a suitable variety of rope sizes and constructions
(c) be able to indicate broken wires and corrosion which reduces the cross sectional area by, say, 2-3%
(d) indicate the distance, in metres, along the wire where the various breaks and types of corrosion occur
(e) capable of use at 0.5-1 m/s.

It would seem that such a device, if available, could be used for routine inspection of mooring cables at the same time as the under-pressure lubrication mentioned is being carried out.

38. Mr Hawkins expresses the view that polyester fibre hawsers should be selected in preference to nylon in view of

their superior wet fatigue life. We would point out that there are other factors which have also to be taken into account when a hawser material is selected. Elasticity is an important matter and for a given strength, polyester is far less elastic than nylon. In order, therefore, to achieve the desired spring stiffness, a polyester hawser has to be significantly longer than a nylon one. When considering the mooring terminal as a whole, this has undesirable side effects, especially an increase in length of the connecting hoses. Another important consideration is the melting point, since heat is produced during the cyclic loading, particularly at the splice and thimble. The significantly lower melting point of polyester and polypropylene is felt to be a disadvantage and would have to be quantified.

39. Recent research carried out on nylon hawsers in California has confirmed that wetting does reduce the fatigue strength of nylon braid line hawsers by about 20%. Other factors, such as rope construction, the effects of overload, and assumptions regarding load sharing between core and sheath are also important. New criteria for determining allowable operating loads for hawsers have been developed and these are being incorporated in a Draft OCIMF Standard. The incidence of hawser failure at terminals where these new criteria have been adopted has decreased significantly.

40. With regard to the points raised by Dr Bowers, we would point out that the mooring pattern analysis in Paper 4 is a simplified one intended to give insight into the characteristics of moorings and methods of calculation. When an engineer is giving consideration to the calculation methods he intends to use for a particular application, he must use engineering judgement when deciding what level of investigation is required and estimating the inherent conservatism or unconservatism of the method he wishes to use.

41. The procedure that has usually been adopted for the design of permanent moorings has been to size the mooring initially on the basis of the method given in Paper 4. An allowance for low-frequency motions may be added at this stage if it is considered necessary based on previous model tests of similar units.

42. For detailed design, a long-duration time domain simulation is generally performed by carrying out model tests in irregular seas with a constant applied wind force equal to the 1 min mean value. Current forces are also applied externally. The load and excursion maxima are then determined from the response spectrum thus established. The model test

wave spectrum often contains a 100 year wave as an additional check on the result.

43. As a further refinement, we would advocate including in the simulation dynamic wind forces also, since, especially in the case of semi-submersibles, they are relatively much more important than second-order wave forces. The second-order wind forces could be developed from a Fourier analysis of wind spectra obtained from real time records such as those contained in reference 5 of Paper 4. The appropriate mean wind speed would, in this case, be the mean hourly value and the drag should be determined from wind tunnel tests and not from codes such as British Standard CP3 which are too conservative for these large structures. Wind effects determined from frequency domain simulations are likely to be unnecessarily conservative and should be avoided. A further possible refinement is the performance of towing tests or calculations to determine the effect of wave and current interaction. Although scale effects are usually not very important, forces and motions determined from model tests should, where possible, also be checked by mathematical models. It should be noted that there are significant differences between waves modelled in a basin and real seas. The mathematical model can thus increase confidence in the results and permit investigation of sensitivity to changes arising during the detailed design of the unit. If there are very significant changes in the configuration, it may then be deemed desirable to carry out further model tests. Finally, the design loads would again be determined using the approach outlined in the Paper but using the mean hourly wind force, combined mean current and wave force and a maximum excursion determined statistically or spectrally from the full band width of the response spectrum which has been developed.

44. Dr Bowers' point regarding the effects of the softening of the mooring due to line breakage is a valid one; however, in the analysis in the Paper, the increased dynamic excursion is counter-balanced by an increased conservatism in the amount of mean excursion due to the use of the 1 min mean wind speed in the damaged condition in place of the mean hourly value. The effects of mooring stiffness on low-frequency motion are investigated in a paper by J. A. Pinkster.[5] In that paper, it is suggested that the RMS of the low-frequency motion depends on the 3/4 power of the mooring stiffness. Considering the stiffnesses at the mean environmental load in Fig. 4 of Paper 4, breakage at one line

would increase the RMS of the second-order excitations by
approximately 25% on that basis.

REFERENCES
1. KITZINGER F. and WINT G. Test results with the Magnograph
   wire rope tester. First Annual Wire Rope Symposium,
   Washington State University, 1980.
2. Interbranch Laboratory for wire rope testing and rope
   transport equipment. Magnetic testing of steel wire ropes,
   University of Mining and Metallurgy of Krakow, 1979.
3. EGEN R. A. Tension member technology: non destructive
   testing of wire rope. Offshore Technology Conference, OTC
   2926, 1979.
4. MARCHENT B. G. An instrument for the non destructive
   testing of wire ropes. Systems Technol., 1978, Aug., No.
   29.
5. PINKSTER J. A. Low-frequency phenomena associated with
   vessels moored at sea. SPE 4387, 1974.

5 State-of-the-art assessment of high capacity sea floor anchors

N. D. ALBERTSEN, MSCE, RCE, and R. M. BEARD, MSE, RCE,
Naval Civil Engineering Laboratory, Port Hueneme, Ca, USA

SYNOPSIS. High capacity anchors of the deadweight, drag
embedment, embedded plate and pile types are discussed in
terms of their physical characteristics, site survey
requirements, potential capacity, advantages, disavantages,
and application. An assessment recent anchor technology
developments is given. Projections are provided on where
anchor technology advancements can be expected to occur
in the near future.

INTRODUCTION/BACKGROUND
1. In 1965 the U.S. Naval Civil Engineering Laboratory at
Port Hueneme, California, began a research and development
program funded by the U.S. Naval Facilities Engineering
Command (NAVFAC) to achieve a deep ocean anchoring capability. Early work involved development of propellant embedded
anchors (PEAs). An example is a 100 kip capacity PEA developed for the Chesapeake Division of NAVFAC. To date, more
than 150 PEA's have been used in U.S. Navy construction
efforts.
2. Beginning in 1978 NAVFAC sought to obtain a better understanding of drag embedment anchors performance. At first
the main interest was in Navy anchors such as the stockless
and STATO types used routinely in fleet moorings. However,
new commercial high capacity drag anchors were included in
the testing because of their potential for Navy use.
3. NCEL has assisted other U.S. government agencies with
anchor related problems. From 1975 to 1977, NCEL provided
assistance to the U.S. Energy Research and Development Agency
in evaluating anchor concepts for the Ocean Thermal Energy
Conversion (OTEC) power plant. This work lead to a fuller
view of the design procedures, limits and applications for
a variety of anchor types including deadweight and pile
anchors.

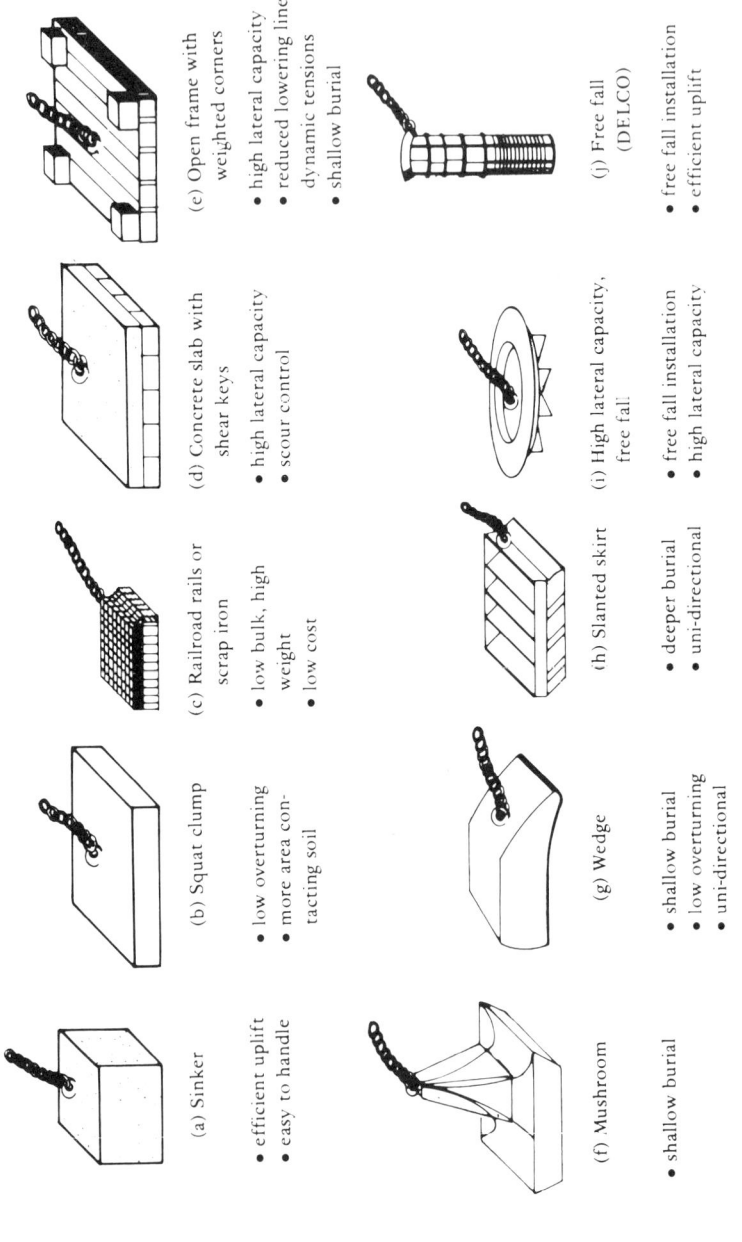

Fig. 1. Several variations on the deadweight anchor

(a) Sinker
- efficient uplift
- easy to handle

(b) Squat clump
- low overturning
- more area contacting soil

(c) Railroad rails or scrap iron
- low bulk, high weight
- low cost

(d) Concrete slab with shear keys
- high lateral capacity
- scour control

(e) Open frame with weighted corners
- high lateral capacity
- reduced lowering line dynamic tensions
- shallow burial

(f) Mushroom
- shallow burial

(g) Wedge
- shallow burial
- low overturning
- uni-directional

(h) Slanted skirt
- deeper burial
- uni-directional

(i) High lateral capacity, free fall
- free fall installation
- high lateral capacity

(j) Free fall (DELCO)
- free fall installation
- efficient uplift

4. Finally, recent experience in constructing pile supported Naval facilities in calcareous materials has shown a need for a better understanding of how piles interact with this material. An implication of this experience is that pile anchors in calcareous soils will not perform as expected. NAVFAC is supporting research to devise methods to improve pile capacity in calcareous soils.

SCOPE

5. This paper summarizes work done by NCEL on deadweight, drag embedment, embedded plate and pile anchors. It also provides a summary of the latest technoglogy developments and gives projections of future advancements.

STATE-OF-THE-ART ASSESSMENT

Deadweight Anchors

6. Deadweight anchors are anchors that depend primarily on their own weight to resist loads. As used here, deadweight anchors include not only anchors that rest on the seafloor but also some anchor types that may be partially or even completely buried in the seafloor.

7. The primary advantages of deadweight anchors stem from their ability to resist uplift forces. This permits the use of shorter mooring scopes and reduces the area required for a mooring, thus enabling higher mooring density. In addition, deadweights can be effectively used on relatively thin sediment layers where pile, drag and embedded plate anchors are not appropriate.

8. The disadvantages of deadweight anchors are their relatively large size and weight compared to other anchor types. Transportation and installation of high capacity deadweight anchors require large crane barges and other heavy load handling equipment either to place a unit anchor or a box-structure for later ballasting.

9. Deadweight anchors vary from sophisticated concrete or steel constructions to engine blocks and concrete clumps. A few typical anchor configurations are shown in Fig. 1. Design procedures for deadweight anchors are well established and a summary is provided in Ref. 1. Site survey requirements include identification of sediment type, thickness and variability, and estimates of soil cohesion, friction angle and scour potential. In general, the capacity of deadweight anchors is limited only by the size and weight limitations imposed by anchor fabrication, transport and installation.

Pile Anchors

10. Pile anchors develop their holding capacity primarily by mobilizing the lateral earth pressure and skin friction in

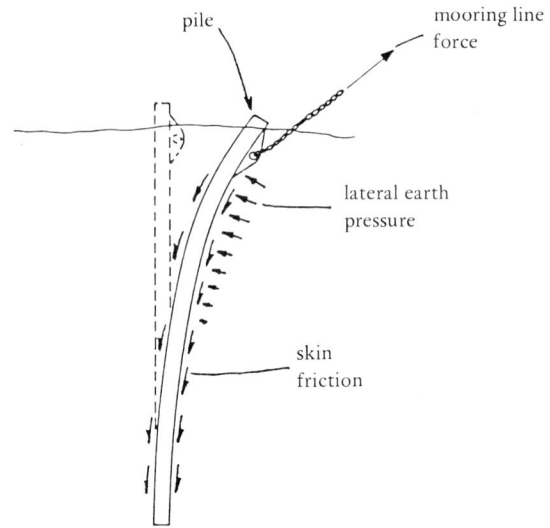

Fig. 2. Skin friction and lateral earth pressure on an anchor pile

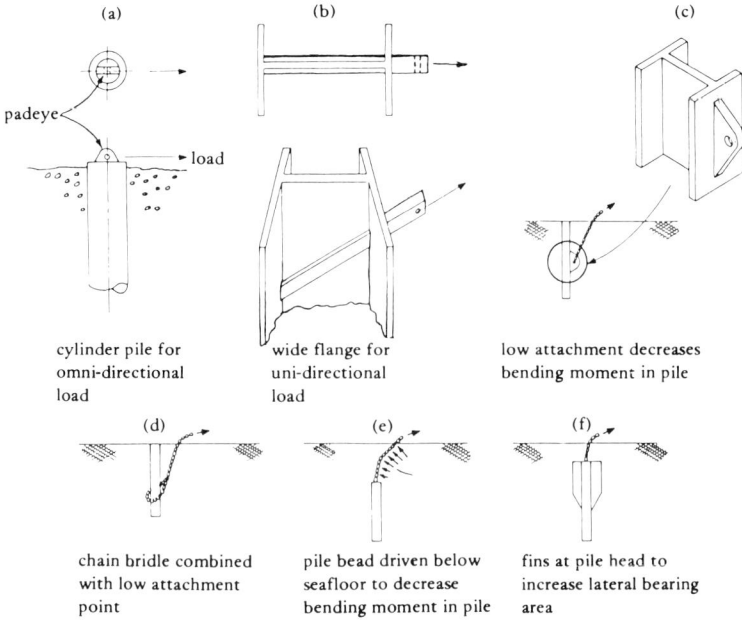

Fig. 3. Variations on the basic pile anchor

the surrounding seafloor material as illustrated in Fig. 2.
11. Their principle advantages are a potential for high
capacity and an ability to resist both lateral and uplift
loads permitting use with short scope moorings. With appropriate drilling techniques they also can be used in
coral or rock seafloors. Their main disadvantages are that
they require specialized installation equipment, costs increase
rapidly in deeper water or in exposed locations, and a
comprehensive site survey is required.
12. Pile configurations vary greatly and are dependent on
water depth, application, seafloor type, loading conditions,
and other factors. Some different pile configurations are
shown in Fig. 3. Pile anchor design involves, for all but
the most simplified cases, a relatively complex multistep
design procedure. A number of procedures are available in
the open literature (Refs. 2 through 5).
13. Site survey requirements are related to the fact that
pile anchors are deeply embedded and the design process
requires detailed geotechnical data to full pile depths.
Required soil properties include strength, sensitivity,
grain size, origin, and density. Since a pile may penetrate
to many tens, and potentially hundreds of feet, acquiring
geotechnical data to comparable depths is difficult and
therefore, expensive.
14. Lateral pile anchor capacity in sediment seafloors is
limited in long pile anchors by bending moments and in short
pile anchors by pile rotation and soil failure. Vertical
capacity is limited by pile friction. In rock seafloors
load capacity is primarily limited by installation technology.
Pile anchor systems capable of resisting mooring tensions of
tens of thousands of kips are thought to be feasible in rock
seafloors (Ref. 6).

Drag Embedment Anchors

15. Drag embedment anchors generate most of their holding
capacity by mobilizing the shear strength of the soil in
which they are embedded. These anchors vary from low efficiency shallow burial,temporary mooring anchors to high
efficiency, deep burial, permanent mooring anchors.
16. Drag embedment anchor characteristics have remained
relatively unchanged for centuries. Their principle advantages are the broad ranges of working experience and anchor
types and sizes available. Also, drag anchors will often
continue to provide near maximum load resisting capability
with drag even though the anchor continues to slip in the
soil. Disadvantages of drag embedment anchors are; they do
not function on rock seafloors, behavior is erratic in layered seafloors, capacity decreases as the pull angle in-

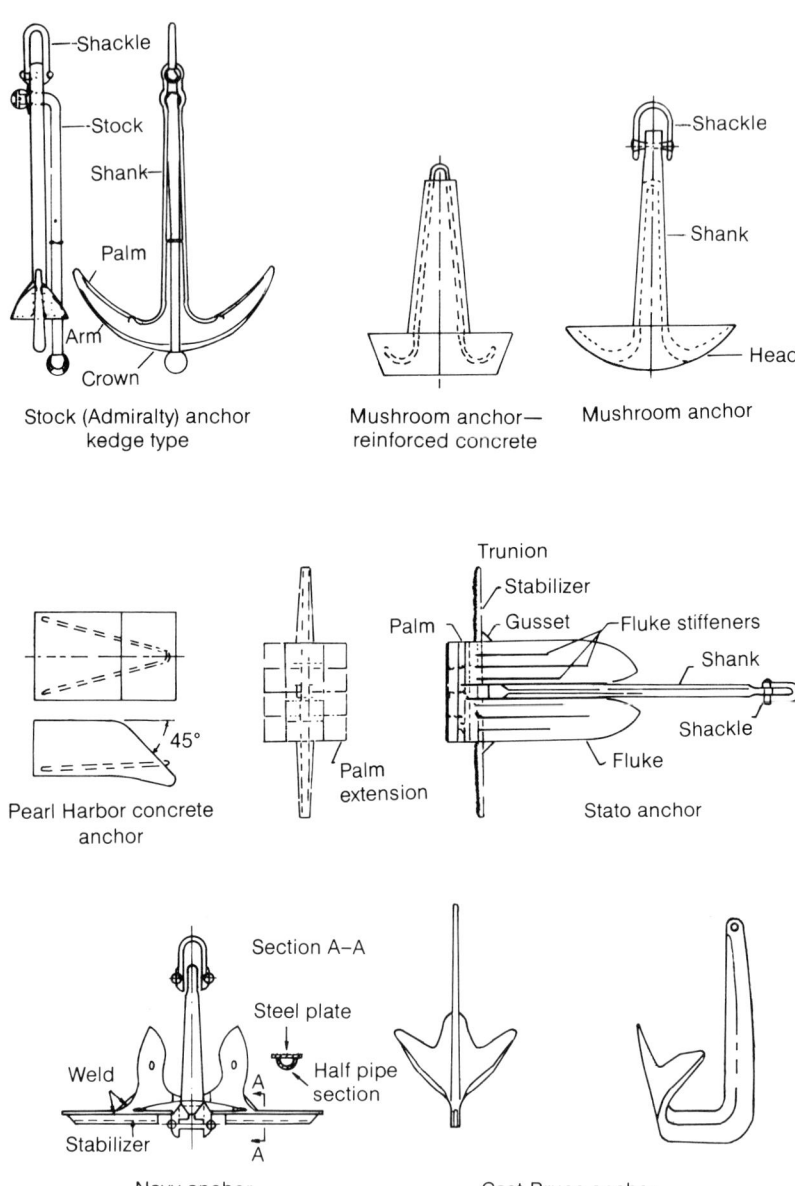

Fig. 4. Examples of drag embedment anchor configurations

creases from horizontal, and they require dragging for embedment and load resisting capability. Dragging can damage
seafloor or subseafloor constructions such as pipelines
or cables. Drag anchors that illustrate a range of configurations are shown in Fig. 4.
17. The use of drag anchors to satisfy mooring requirements
poses an anchor selection problem rather than a rigorous
design problem. Usually a selection relies on engineering
judgment formed from previous experience or manufacturer
claims. To help relieve the uncertainity in the anchor
selection process, the Navy has been conducting an anchor
test program (Refs. 7, 8, 9, and 10). Table 1 is a summary
of data developed recently for several types of U.S. Navy
and commercial drag embedment anchors at four locations
typical of Navy harbors. In Table 1, efficiency is the ratio
of anchor holding capacity measured at the anchor to actual
anchor weight. The reader is cautioned that the data in
Table 1 are site specific; anchor performance could vary for
other seafloor conditions. Inferences from the data should
only be drawn after the site conditions and test records are
reviewed.

Table 1. Efficiencies of Drag Embedment Anchors in Two Types
of Seafloors.

ANCHOR TYPE	ANCHOR WEIGHT (LBS)		EFFICIENCY	
	NOMIMAL	ACTUAL	SAND	MUD/SILT
Navy Stockless	5,000	5,950	8:1	3:1
STATO	3,000	3,500	18:1	15:1
Stevfix	1,410	1,410	31:1	15:1
Stevdig	2,200	2,560	29:1	-
Stevmud	2,200	2,200	-	20:1
Hook	1,230	1,260	12:1	18:1
Bruce	1,320	1,320	25:1	-
Bruce Twin Shank	750	750	-	12:1

Embedded Plate Anchors
18. Plate anchors are identified by the way they are embedded
into the seafloor. The embedment methods include augering,
vibrating, jetting, hammering, and shooting by a propellant
charge.
19. Embedded plate anchor advantages are the ability to
resist uplift loads enabling short scope moorings, high
holding capacity to weight ratios, use in layered seafloors,
and accurate placement. The disadvantages are that in many
cases specialized equipment are required to embed the plates,

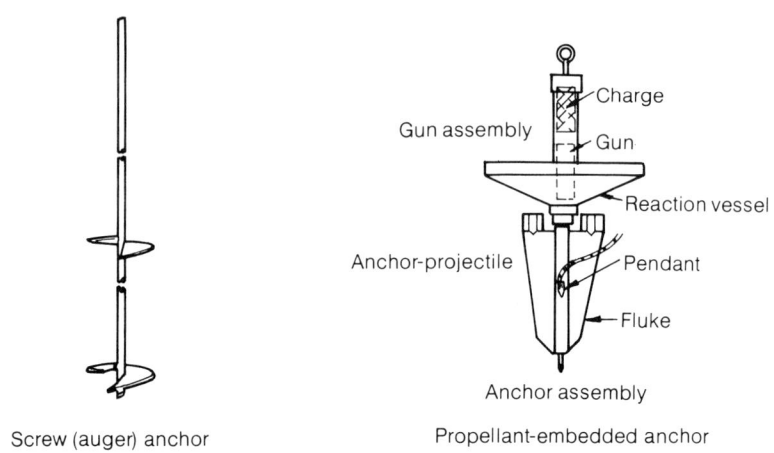

Screw (auger) anchor

Propellant-embedded anchor

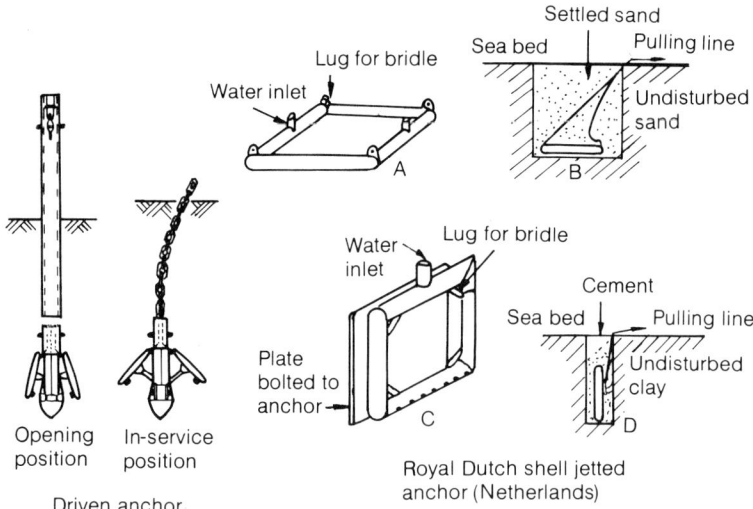

Opening position   In-service position

Driven anchor

Royal Dutch shell jetted anchor (Netherlands)

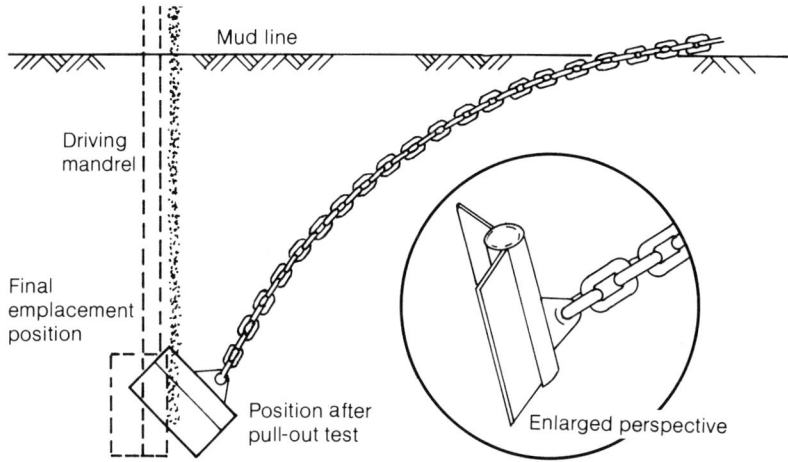

Menard rotating plate anchor

Fig. 5   Examples of embedded plate anchors

they are susceptable to cycle load strength reduction in loose sand and coarse silt, and the plates are generally not recoverable. Other disadvantages are that once the plate holding capacity has been exceeded or plate movement has occurred, holding capacity is reduced and that use is generally limited to sediment seafloors (the exception being propellant driven types). Some embedded plate anchors are illustrated in Fig. 5.
20. Since embedded plate anchor holding capacity is influenced by loading condition and soil type, the design or selection of a plate anchor should be preceded by an evaluation of the conditions of anchor operation. The first step should be a geotechnical site survey. For non-critical installations depth and type of sediment may be all that are required. For critical installations geotechnical parameters of importance are strength, sensitivity, density, plasticity, grain size distribution, permeability, origin, and depth to rock. Once this data is available, the capacity of a plate can be determined as a function of depth for the loading conditions expected (Ref. 11).

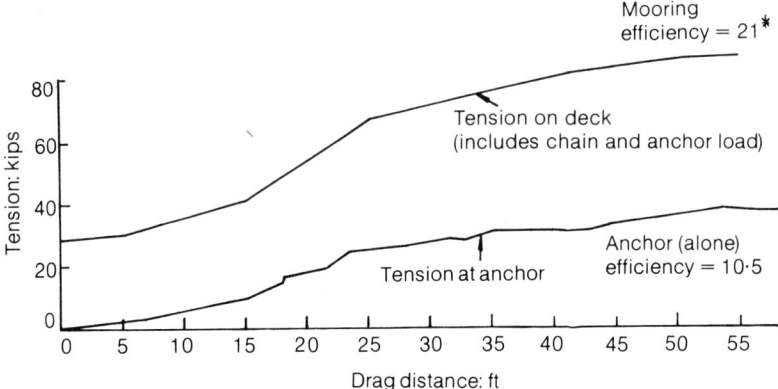

Fig. 6. Load against distance for 3000 lb STATO anchor showing effect of chain in mud (*chain drag on bottom subtracted from deck tension to calculate efficiency)

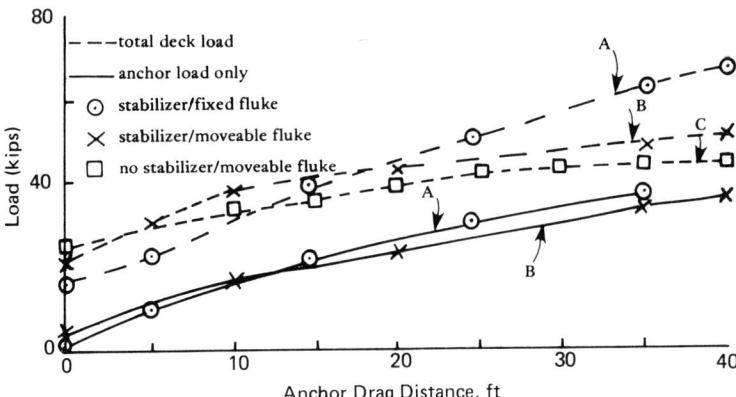

Fig. 7. Load against distance for 9000 lb Stockless anchor showing effect of chain in mud

RECENT DEVELOPMENTS
Deadweight Anchors
21.   When conducting the OTEC anchor concept study it was
noted that while deadweight anchors could be rationally designed, it required a number of seemingly unrelated analysis
techniques.   Ref. 1 has brought these methods together into
a systematic design procedure. With this procedure more
efficient use of deadweights can be achieved.
Pile Anchors
22.   Recent U.S. Navy experience has shown that piles driven
into calcareous materials do not consistently behave according to any known design theory.   In one instance, piling
requirements were three times that indicated by state-of-
the-art procedures.   This problem is believed to result from
a lack of side friction development caused by grain crushing
and cementation working to prevent development of significant
lateral pressure on the piles.   There is also the possibility
of a boundary layer of crushed material with poor frictional
characteristics between the pile and the hole.   This poor
frictional resistance to support piling is also a problem
for pile anchors.   Pile anchors in calcareous materials
designed according to present practice would probably be
under capacity.   To overcome this problem, a series of laboratory tests have been undertaken to more clearly define
the roles of grain crushing, fraction of crush material, and
degree of cementation in friction pile capacity in calcareous
soils.   The tests are just beginning, but have already shown
that even light cementation can sharply reduce frictional
resistance.
Drag Embedment Anchors
23.   Data obtained from recent tests of U.S. Navy and commercial high capacity anchors are being used to develop a better
understanding of the role anchors and anchor chain play in
mooring efficiency (capacity of the anchor plus the holding
capacity of the embedded chain divided by anchor weight).
The test data show that the embedded anchor chain produces
a significant amount of added holding capacity.   This behavior is shown in Figs. 6 and 7 for several different
anchors in mud and sand. For the case of the 3,000 pound
STATO anchor in mud, Fig. 6, it can be seen that more than
50 percent of the holding capacity is produced by the embedded chain.   The magnitude of this effect changed with changes
in anchor and chain tested.
24.   Other general observations that the data support are:
(1) Small anchors (<1,000 lb) often exhibit higher efficiencies than anchors of 10,000 to 30,000 lbs.   (2) Anchor roll
stability is very sensitive to stabilizer design; poor

performance can sometimes be corrected by extending stabi-
lizers. (3) The effect of mooring line angle at the seafloor
has a major impact on anchor holding capacity; as the angle
increases anchor holding capacity decreases. (4) The angle
between anchor fluke and shank has a major influence on
anchor holding capacity; for mud the optimum angle is 50°,
for sand the optimum angle is 30° to 35°. (5) While the type
of mooring line used (wire vs chain) does not have significant impact on mooring efficiency it is clear that when
chain mooring lines are used, anchor penetration and drag
distance to peak load are less than with cable mooring lines.
(6) Conventional drag embedment anchors can be effectively
used in tandem; the most efficient rigging for tanden anchors
is to attach the chain from the rear anchor to the shank
shackle of the front anchor.

Propellant Embedded Anchors

25. Two recent developments have occurred with the propellant
embedded anchors. The first involves a PEA system with a
nominal holding capacity of 300 kips. The system has been
fabricated and test fired; however, some structural problems
in the reaction vessel must be overcome before this system
can be considered for routine use.

26. The second development centers on the work being done
with rock flukes for the PEAs. Tests of different fluke
shapes indicated that a conical shape has the highest potential for success in rock. A number of these flukes were
fabricated and fired into sandstone and vesicular basalt
using the 20K PEA system. The results of these tests
have been very encouraging; holding capacities of more than
60 kips have been consistently achieved. In the first pull
tests the pull cables failed without moving the fluke. Actual
capacities will be determined when the flukes are retested.

TECHNOLOGY ASSESSMENT/PROJECTIONS

27. The design of pile anchors is for the most part an
advanced technology utilizing the procedures for designing
laterally loaded and friction piles. An exception is the
design of pile anchors in calcareous soils. Advances in
this design area are expected that will allow for better,
more accurate designs and hence reduced costs.

28. The next few years will see major advances in procedures
for selecting and sizing the drag embedment anchors. Work
now being done to develop mathematical models of anchor and
chain behavior as function of soil engineering properties
will lead to rational selection procedures for drag anchors.
This will be a significant improvement over present methods

of extrapolating the results of small anchor tests. These advances may result in more efficient anchors and anchor system designs.

29. PEA technology will continue to progress in the area of size. It is anticipated that the 300K anchor under development will provide for capacity up to 600 kips in sand seafloors. Recent use of PEAs in rock seafloors is highly encouraging and should lead to the capability being developed for the large PEAs. The holding capacity potential in rock for a given size anchor should be established and is expected to be many times the nominal capacity of the anchor.

SUMMARY

30. NCEL has worked on anchor technology for more than a decade. The state-of-the-art of deadweight, pile, drag, and embedded plate anchors allows for their use in many anchoring situations. The technology of each type has been advanced in recent years. Continued advancements in both hardware and design areas are expected for pile, drag, and embedded plate anchors.

REFERENCES

1. Valent, P.J., R.J. Taylor, H.J. Lee, and R.D. Rail. State-of-the-art in high capacity, deep water anchor systems. Civil Engineering Laboratory, Technical Memorandum 42-76-1. Port Hueneme, CA, January 1976.

2. Valent, P.J., R.J. Taylor, J.M. Atturio, and R.M. Beard. Single anchor holding capacities for ocean thermal energy conversion (OTEC) in typical deep sea sediments. Ocean Engineering, Vol. 6, Nos. 1/2, Pergamon Press Ltd. Oxford, England, 1979, pp 169-245.

3. McClelland, B. Design of deep penetration piles for ocean structures. Journal of the Geotechnical Engineering Division, American Society of Civil Engineers. Vol. 10, GT7, July 1974, pp 704-747.

4. Focht, J.A. and F.J. Koch. Rational analysis of the lateral performance of offshore pile groups. OTC 1896. 1973 Offshore Technology Conference, Preprints, Vol. II, Houston, TX, pp 701-708.

5. Matlock, H. and L.C. Reese. Generalized solutions for laterally loaded piles. Journal of the Soil Mechanics and Foundations Division, American Society of Civil Engineers, Vol. 86, No. SM5, Part I, October 1960, pp 63-91.

6. Valent, P.J. and J.M. Atturio. OTEC anchors: selection and plan for development. Civil Engineering Laboratory, Technical Report R-859, Port Hueneme, CA, December 1977.

7. Taylor, R.J. Performance of conventional anchors.
OTC 4048, 1981 Offshore Technology Conference, Preprints,
Vol.II, Houston, TX, May 1981, pp 363-372.
8. Taylor, R.J. Test data summary for commerically avail-
able drag embedment anchors. Report for official use only,
Civil Engineering Laboratory, Port Hueneme, CA, June 1980.
9. Taylor R.J. Conventional anchor tests results at San
Diego and Indian Island. Civil Engineering Laboratory, Tech-
nical Note N-1581, Port Hueneme, CA, July 1980.
10. Taylor, R.J. Conventional anchor test results at Guam.
Civil Engineering Laboratory, Technical Note N-1592, Port
Hueneme, CA, October 1980.
11. Beard, R.M. Holding capacity of plate anchors. Civil
Engineering Laboratory, Technical Report R-882, Port Hueneme,
CA, October 1980.

<u>CONVERSION TABLE</u>
Feet x .305 = meters
Pounds x .455 = kilograms
Kips x 4.45 = kilonewtons

6 Offshore mooring system approval for insurance purposes

S. M. TAIT, BSc, MICE, London Offshore Consultants Ltd

SYNOPSIS. Most major offshore structures are insured during their construction and installation phase. It is normal for the terms of the insurance to require an independent check on marine operations by a technical organisation, nominated by the Underwriter. Mooring systems are included within this area of approval. It is likely that moorings for the offshore installation of fixed and floating production system will become more numerous over the next few years. This paper is intended to act as a guidance document to assist those mooring designers and contractors who may be required to submit their proposals for insurance approval.

INTRODUCTION

1. Once the insurance risk has been accepted the assured will be offered the names of two or three independent technical organisations. He can select one to act as Marine Insurance Surveyor on his project. The Surveyor checks and approves the design and procedures developed for all marine operations, during the construction and installation phase of the structure. These are prepared by the Operators nominated designer and offshore installation contractor and are submitted to the Surveyor in the early stages of development of the project. It may be that the Operator also appoints a Surveyor in a consultancy role to assess the feasibility of a new concept prior to major contractual commitments being made.

2. Inshore and offshore moorings often represent an important phase of marine operation and the Surveyor will normally issue a seperate Certificate of Approval for the mooring system once it has been satisfactorily designed and installed and prior to the structure being connected. The documentation required is no more or no less than that which should, in any case, be prepared by designers and contractors for their

clients. It maybe that the Surveyor has, because of his involvement on many projects, greater experience of the proposed mooring system. In which case he should adopt a "constructive approval" role and assist all parties in producing the most effective and cost efficient system. This paper although referring primarily, to the insurance approval aspect should be considered as a general guidance document on the proper design and execution of a mooring system. Although the mooring systems of semi-subs, crane ships etc. may require insurance approval, it is not the purpose of this paper to refer to such moorings, except when they are associated and integrated, with the installation of a major offshore structure.

MOORING SYSTEMS REQUIRING APPROVAL

3. Examples of mooring systems which will typically require approval, as being part of the installation of an offshore structure are as follows:

a) Gravity Platform Construction Moorings
These medium term (1-2 year) moorings are for the construction of gravity platforms of steel or concrete and are usually installed in a Norwegian Fjord or Scottish Loch.

b) Gravity Platform Installation Moorings
These moorings are for the offshore exact emplacement of a gravity structure over a pre-installed template and could be described as short term (for operations lasting a few hours or days).

c) Deck Mating Moorings
These are short term moorings for the installation of deck segments or fully intergrated decks inshore on gravity structures or other types of floating structure. They also include moorings for any marine operation where a structure is to be integrated during the construction phase afloat.

d) Deck Installation Moorings Offshore
These moorings are for the offshore installation of an integrated deck structure from a barge onto a fixed platform such as a piled jacket.

e) General Offshore Moorings
For the installation of Guyed Towers, Jack-up Production Platforms over Templates, Articulated Structures, Wave Energy Devices etc. These would include construction vessels **where their mooring system are** critical to the operation.

144

f) Floating Production System Moorings
These are long term (5-25 year) moorings which secure the floating production system or T.L.P. throughout its operational life and where the permanent system is partly or wholly used to secure the structure immediately it arrives at the field.

SUBMISSIONS TO BE MADE

4. Depending on the nature and scale of the project and the exact wording of the insurance policy, it is probable that the marine insurance surveyor would be seeking submission by the designers and contractors of documentation procedures and calculations on a) mooring designs b) procedures

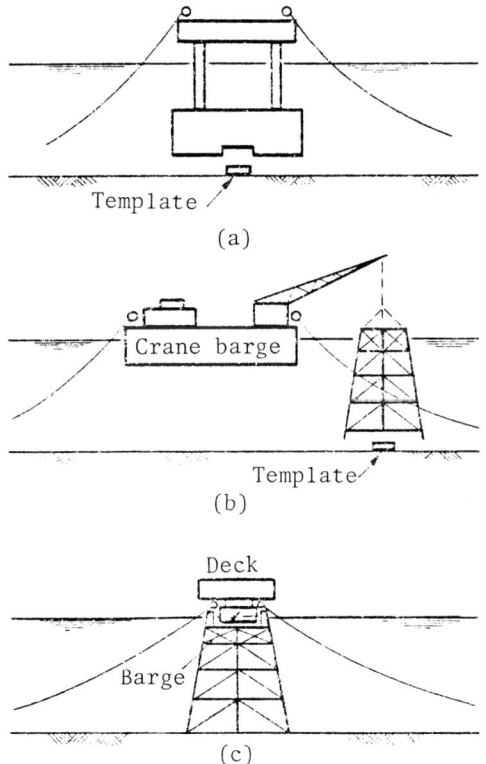

Template

(a)

Crane barge

Template

(b)

Deck

Barge

(c)

Fig. 1. Some offshore mooring operations: (a) exact emplacement of gravity structure; (b) exact emplacement of steel jacket; (c) installation of integrated deck structure

5. MOORING DESIGN DETAILS

a) Environmental data for the location (measured or predicted)
b) Hydrographic data including updated charts and surveys
c) Seabed soils data of mooring area including borehole results at proposed anchor locations.
d) Drawings of the mooring structure including temporary construction equipment and marine floating plant.
e) Drawing and details of seperate mooring systems of marine construction equipment.
f) Construction and/or installation programme indicating required life of mooring, intended period of the year for installation and operation, contingency plans for delay, (eg. if moorings are designed for Summer conditions what are plans for Winter?).
g) Calculations of environmental forces on structure and other equipment.
h) Calculations of mooring line tensions or defined mooring configurations and evaluation of component parts with defined safety factors.
i) Calculations of excursions of the moored structure and comparison with the allowable.
j) Calculations of component parts such as anchors, winches, stoppers, fairleads etc.
k) Detailed and general drawings of the total mooring system.

6. PROCEDURES

a) Test and inspection procedures for component parts.
b) Mooring installation procedure developed by marine contractors including preparations for connection.
c) Anchor installation procedure.
d) Mooring tension procedure
e) Structure connection procedure
f) Pre-tensioning procedures, if applicable.
g) Operational procedure if moorings are actively used in an installation operation (eg. exact emplacement over a template)
h) Maintenance and repair procedure if moorings are of medium term duration.
i) Disconnection and demobilisation procedure if it is critical to the security of the insured structure.

7. ATTENDANCES

In addition to the review of documentation the surveyor will possibly require to be represented by one of his Master Mariners or Engineers (according to the operation) at various operations as follows:-

a) Tests of component parts eg. anchor trials,
b) Inspection of parts to be reused.
c) Mooring installation operation.
d) Structure connection operation) Attendance may be
e) Installation Operation   ) required with the terms
f) Disconnection Operation   ) of the towage warranty.

SYSTEM SELECTION

8. Prior to choosing the most desirable form of mooring
system, it is necessary to clearly define the operational
requirements. These requirements should be based on:-
duration of the operation; installation tolerances;
exposure of the structure and of the mooring system; required
degree of redundancy; allowable excursion; speed and
requirements for installation and connection onto the mooring;
cost benefits to be obtained in refining design eg. by obtaining
good soils data good environmental data and weather data etc.
Once the mooring requirement has been established, then the
alternative forms of mooring system could be considered.
The following is a list which represents a sample of common
mooring systems used today.

9. Catenary Moorings - the simple catenary is the most
common form of mooring arrangement which has been around
for many years. It often involves the combination of chain
and/or wire and drag-in anchors. It can be used in single
leg or multi leg systems and it relies on the gravitational
forces of the suspended rode material to resist horizontal
movement.

10. Catenary plus Clump - this arrangement is similar to the
simple catenary but involves the addition of clum weights
of steel orconcrete on a portion of the suspended rode.

(Note: the term "rode" describes the connecting material
in a mooring, which maybe chain, wire, polyprop, nylon,etc).

11. Taut Line Moorings - are much shorter in length than
Catenary and involve lighter, more elastic materials such as
wire ropes and man-made fibres. They utilise the extensible
properties of flexible rode to provide the restoring forces.

12. Taut line and Submerged Buoys - by introducing buoys
into a mooring line, it is possible to produce a situation
similar to inverted catenary and clumps. The buoys provide
steady uplift forces when totally submerged, thus adding to
the elasticity of the system.

13. Tension Buoy - this taut line system involves the use of a single buoy located above the anchor and tensioned against excess buoyancy towards it. A resulting form of inverted pendulum then gives resistance to horizontal movement when residual buoyancy remains in the buoy.

14. Tension Devices - mechinical and gravitational facilities can be introduced on the moored structure to provide resistance to applied steady and dynamic forces. Springs or hydraulic dampers can be used to obsorb all or part of the energy in the rode, or the rode can be run over pulleys such that a lump of steel or concrete is suspended below the water line.

15. Tension Leg System - Tension leg systems are similar to Tension Buoy systems in that lateral resistance is developed by generating excess buoyancy. The difference being in that the tension leg system the moored structure itself is held in tension against its own buoyancy.

16. Dynamic Positioning Systems - these are automatically controlled propeller systems based on a sophisticated navigational unit. This method is an active system and frequently adopted for deep water temporary mooring situations on purpose built vessels.

17. Yoke Moorings - these are usually vertical articulated steel tower or frame moorings secured to a gravity base on the seabed and are used for mooring tankers for offshore storage purpose.

MOORING DESIGN

18. Mooring Layout - In designing mooring system one of the most difficult problems is deciding the necessary degree of redundancy. Mooring systems normally consist of legs numbering 1 to 16 according to the requirement for restricted excursion, degree of redundancy, scale of forces etc. The legs are usually equally distributed around the floating body. High leg numbers, whilst giving lower forces on the legs and a greater degree of redundancy, create problems in equal distribution of forces.

19. It is desirable that the mooring systems have an adequate degree of redundancy such that failure of one leg will not lead to loss or extensive damage of the structure. It can be shown that 3 and 4 leg mooring systems do not have an adequate degree of redundancy in many cases.

20. In choosing the number of mooring legs, the following should be considered:-
a) Maximum capacity of component parts.
b) Seabed topography in relation to maximum draft of the

floating structure.

c) Implications of the structure running aground if the allowable excursions are exceeded.

d) The environmental forces and direction of wind, wave and current.

e) Implications of the structure making contact with a seabed structure eg. preinstalled seabed template or pipeline.

f) Increase in forces on those mooring components which remain intact if failure in one leg occurs.

DESIGN LOADS

21. Once a mooring configuration has been determined, detailed design can proceed and the following loads should be considered:-

Dead Loads - on the mooring system including loads from weight of components, buoyancy of the system and buoyancy of the moored structure.

Imposed Loads - from equipment which can be installed on the structure during the period that it is at the moorings. Also changes in buoyancy of the structure and forces from ships and other floating vessels moored to it.

Environmental Loads - from wind, wave and current in particular (plus ice and earthquakes forces if applicable). The environmental loads usually predominate in the design of a mooring system.

ENVIRONMENTAL LOADS

22. Wave Loads - Wave loads dominate for those structures which are of shallow draft and low height above the water. They should be determined by computer simulation, assisted where ever possible, by model tests. The calculations should be based on defined criteria based on wave data predicted by a reliable authority. The importance of obtaining reliable and accurate wave data over the longest possible period, cannot be understated. The directionality of the wave should also be obtained as this can lead to more efficient mooring design. Normally, significant wave height is used in mooring design and most insurance approval organisation adopt a criteria of 100 year return period for medium term moorings and 10 year for short term moorings.

23. Wind Loads - the wind loads should be determined for all parts of the structure above the mean water line. They should include temporary construction and maintenance equipment (frequently temporary equipment is ignored and this

can add considerably to the resistance area of the structure). Standard formulae are available for calculating wind forces. The wind speed should be based on accurate predicted values based on recorded figures for the location as close as possible to the mooring position. 10 year return and 100 year return 1 minute mean wind speeds are normally used in the calculations.

24. Current Forces – Current speed should be recorded at the proposed mooring location, at different depths over as long a period as possible, in order to predict possible current forces and directions. Calculation of current forces on the moored structure should be adopted using common formulae.

25. Total Forces – The total environmental forces should then be calculated by the summation of wind, wave and current forces, directionalised as appropriate. Mooring line tensions should be determined by computer calculation for the defined loading conditions for all possible positions, drafts and configurations of the moored structure. A force diagram and excursion envelope should be developed which will define the limits of movement of the structure. This envelope can then super-imposed over the hydrographic plan of the moored location and underkeel and horizontal clearance can be established at maximum excursions. Calculations should be undertaken with one or more mooring legs broken to establish the further limits of excursion and the maximum forces in the unbroken mooring legs. The position of the maximum excursion of the structure in this condition will indicate the change in layout of the mooring pattern. This in turn should indicate whether the anchors are subjected to lateral forces greater than they can withstand.

SEABED SOILS INVESTIGATION

26. Prior to deciding which type of anchor to use, it is necessary to carry out detail seabed investigations. This investigation should include determination of the nature and depth of the seabed material and their properties and should be compatible with those investigations carried out in determining the nature of the foundation material for a fixed structure. Normally, the depth of seabed sampling and penetration tests could be limited to say, 20–25 metres below the seabed because it is unlikely that any form of anchor will penetrate below this depth. Care should be taken to determine the extent and nature of any boulders or rock out-crops in the vicinity of the proposed anchor location, as this could have influence on determining the choice of anchor.

The seabed investigation should be carried out over an area
which should take into account the tolerance of installation
and the possibility that the anchor may need to be recovered
and relaid.

ANCHORS

27. Drag-in (or embedment) Anchors - this is the traditional
type of anchor used primarily in Catenary mooring systems
with chains or wires. These anchors generate resistance by
penetrating into the soft seabed materials by their own
weight and shape. Early designs which have remain unchanged
for hundreds of years relied primarily on their own weight
for resistance. They were limited to not more than 20 tons
for handling reasons. Developments over the past decade
have lead to the introduction of more scientifically designed
anchors. Anchors of this type are currently available in
weights up to 40 tonnes. Drag-in Anchors are best suited for
anchor types which will require recovery in softer seabed
material. They are susceptible to damage by rocks or boulders,
they cannot accommodate directly applied vertical forces and
are not suited to a rocky seabed. Also, they can only accept
horizontal forces over a narrow arc and they draw the adjacent
section of rode material down into the seabed, thus causing
reconnection and corrosion problems. In generating the
horizontal resistance, most forms of anchor move horizontally
as well as vertically, this means that they are not really
suited for conditions where anchor drag is not allowed, eg.
in the vicinity of seabed pipelines. However, most forms of
dra g-in anchor are well proven, they are used for all the
major offshore vessels and have also been used successfully
for medium term moorings, for inshore gravity structure
construction.

28. Gravity Anchors - Gravity or dead weight anchors of
various sizes and types have been used successfully for
medium and long term moorings but are subject to high
installation and removal costs. They provide resistance to
lateral movement by developing a friction force between the
underside of the structure and the seabed which can be
increased by adding skirts or dowells. Large anchors can
lowered into position with cranes or they can be floated.
In theory there is no limit to the size of anchor that can
be installed if it is a self floating and installed type.
They are less dependent than Drag-in anchors on seabed
conditions and they can be omni directional as regards the
applied force.

29. Pile/Anchors - Pile/anchors have not been used extensively offshore because the drag-in anchor has been quite adequate for the applied forces to date and is much less expensive to install. However, pile anchors will be used more offshore in the future, because permanent offshore moorings are required and because there is a need for anchors which will not move under load. Also they are more likely to give proven predictable holding resistance. They are less susceptible to unpredicted seabed conditions which may vary from soft to quite firm geological formations. They can be cement grouted into bored holes or driven or screwed either singularly or in groups linked with the seabed template. Anchor piles in very soft material would have the connection point well down in the pile in order to balance the ground resistance and prevent pile rotation and liftout. Other-wise the rode connection would be above the level of the seabed. This allows ease of connection/disconnection and is less likely to lead to corrosion of the rode materials. Pile anchors offer a high weight to holding power efficiency and they can be omni directional to the applied force and can be designed to withstand vertical uplift forces. They are also more easily linked togehter. They can be expensive if installed in small numbers but become more economical as the numbers are increased beacuse of the spread of pile driving mobilisation costs.

30. Uplift Resistant/Self-burying Anchors and Other Types
There is a group of anchors which can be seperated from the drag-in, gravity and pile anchors, even though in broad terms these main types would come under the description of uplift resistant or self-burying. This include vibratory anchors, screw anchors, propellant activated direct embedment anchors, and plate anchors.

31. General Considerations
In choosing an anchor the following should be considered :-

a) The nature and depth of the seabed material;
b) The maximum and repeditive loads;
c) The extent of anchor drag/deflection/embedment;
d) The influence of cyclic loads and seabed soils;
e) Is anchor drag more acceptable than anchor failure at maximum loads?
f) Do the anchors need to be recovered?
g) Is there heavy marine equipment already on site which can install pile anchors thus reducing mobilisation costs?
h) Do the anchor handling vessels have sufficient deck

space and roller winch capacity to handle large drag-in anchors or are special heavy lift vessels required?

i) Is there a limited weather window for anchor installation? If so, speed of installation is important.

j) Are the loads applied to the anchor likely to spread over large horizontal and vertical arc?

k) Does the rode material adjacent to the anchor need to be replaced for maintenance purposes?

l) What prototype information is available on the chosen anchor?

m) Will model/fullscale test be necessary?

n) Will the anchor require to be pre-tensioned for proper embedment?

RODES

32. Chains - Stud link chains are well proven and traditionally used in Catenary moorings for high loads, medium and long term situations. Chain is made up of seperate components each subject to failure. It is less likely to be used for future offshore short term moorings associated with the installation of a structure except when the installation mooring form part of the permanent system such as with a guyed tower or floating production system. In these cases particular attention should be paid to the programme of installation as it will involve several weather windows.

33. Wire Ropes - Safety factors on wire ropes are generally more conservative than on chains because of increased corrosion and because of strength reductions due to handling, radius bending, fittings etc. For short term offshore installation moorings wire rope has advantages as it is lighter, easier to handle, can be incorporated in winch control systems for exact emplacement and because it will normally allow structure connection within one weather window even with multi leg systems. In choosing a suitable wire rope consideration should be given to:- the lay of the rope; inner core (IWRC); lubrication; outer protection reductions due to bending; oblique loading fatigue; seabed abrasion; crushing; etc.

34. Synthetic Ropes - These can be of nylon, polyester, polypropelene or new materials such as Kevlar. Because of their greater ease of handling they offer considerable potential for short term mooring operations particularly if greater strength is available with parallel filoments.

INSTALLATION AND CONNECTION OPERATIONS

35. The Insurance Surveyor will pay particular attention to the procedures for installation of the mooring system and for the preparatory work associated with structure connection/disconnection. As mentioned previously, he will be represented in the field during these operations. He will be particularly interested in the following :-

a) Survey methods used for anchor and rode positioning.
b) Proof of anchor penetration (TV, Sonar survey etc).
c) Proof of rode emplacement (accurancy of laying etc).
d) Method of anchor/rode recovery if failure occurs (note: it is preferable for all submerged parts to be stronger than easily accessible parts).
e) Buoying off of connections to structure.
f) Methods of pretensioning system, if applicable.
g) Adequacy of spare rodes and fittings for connection.
h) Connection programme within defined weather windows.
i) Winch control and wire tension monitoring systems.
j) Subsea position fixing system for exact emplacement.
k) Disconnection procedures.

CONCLUSIONS

36. Insurance approval of moorings should be constructive rather than destructive. Those organisations given the responsibility to approve the work of others should provide assistance and guidance at all stages of development and be part of a team in the success of an operation. Designers and contractors should consider moorings as an integral part of the project and not ancillary. They require the same level of attention and effort as the permanently installed structure. Authorities and all organisations should be aware that total loss of structure and serious loss of life could occur through failure of a mooring system, particularly during the installation phase, and that there is a real need for positive guidance in mooring design and development compatible with that currently available for fixed structures.

Discussion on Papers 5 and 6

MR W. A. DELL, Rendel Palmer and Tritton

1. Can the Authors of Paper 5 explain the names of the anchor types in Table 1, e.g. Stevfix? In relation to the performance of anchor plus chain against an anchor on its own, how was this measured - at the anchor and chain or at the chain alone (say on a wire)? The performance of a high-holding-type anchor might in fact be inhibited by the presence of the chain. Can the Authors clarify the conclusion arrived at in note (5) of paragraph 24, as this appears to contradict an earlier conclusion in the Paper?

MR P. BRUCE, Bruce International Ltd, The Isle of Man

2. Regarding Paper 5, the tension measured at an anchor shackle (as set out in Table 1) is not necessarily representative of the tension measured at the pulling barge, since the chain on a deep burying anchor will develop higher drag forces in the soil than will occur in the case of a shallow burying anchor. Thus, a large fluke area anchor at shallow depth may give a higher tension at the anchor shackle than an anchor of smaller fluke area at a greater burial depth, but the latter may give greater tension at the barge.

3. The 750 lb Bruce twin-shank anchor mentioned in the table was the first prototype of the Bruce fabricated anchor (tested at Indian Island in Puget Sound). Later tests, for classification society high-holding type approval, performed near Stavanger, Norway, and witnessed by surveyors from DNV, Lloyds, ABS and NKK in November 1981 using the final design of Bruce twin-shank anchor, gave efficiencies measured at the pulling vessel of 41 in sand and 39 in mud. This final design incorporated modifications following suggestions made by Mr R. Taylor of the Naval Civil Engineering Laboratory, Port Hueneme, consequent to the trials with the prototype

anchor.

REAR ADMIRAL A. F. R. WEIR, *G. Maunsell & Partners*

4. Regarding Mr Dell's contribution at the beginning of this discussion, chain is very important from the seaman's practical point of view. It is essential, from my experience in the Royal Navy, for at least the last shackle of a mooring to be able to assist the anchor with holding power and to ensure a horizontal pull at the anchor shackle.

5. To achieve this in the Royal Navy we use cable length equal to at least eight times the depth of water; wire alone would enable a horizontal pull to be maintained. Dredgers usually work in much shallower water where much more than eight times the depth of water of a mooring is easily possible and therefore wire, more easily worked, is an acceptable material.

MR J. D. MILLER, *Cambridge University Engineering Department*

6. The suction pile is an interesting development in offshore anchorage that has recently been deployed in the Gorm Field by single-buoy mooring (SBM). Emplacement of the pile, which is sealed at the upper end, is effected by evacuation, and so does not require fixed driving equipment. However, little work has been reported on the failure or breakout mechanisms.

7. As a final year project at Cambridge University I have undertaken model tests to study piles in saturated loose sand. A model pile was tested with varying position of tether point and angle of pull (Fig. 1). Failure of the pile was determined by the break in slope of a load/log displacement curve. Movement was determined by three dial gauges acting on a gauge arm fixed to the top of the pile, and the displaced positions were drawn using a computer graphics system. The results are illustrated in Fig. 2, which shows that the capacity of the pile is increased by positioning the tether point at a depth of two thirds of the pile length instead of the upper end, and Fig. 3, which shows that for the tether at that depth the vertical capacity is greatest when the angle of pull is $\alpha = 30°$.

8. From Fig. 3 it appears that in calculating the vertical breakout capacity V of the suction-emplaced pile it is appropriate to add the weight of the plug of soil within the cylinder to the weight of the cylinder to obtain a total weight W, and that V≈W. Is this a reliable calculation for a mooring under sustained or cyclic loading, and is it

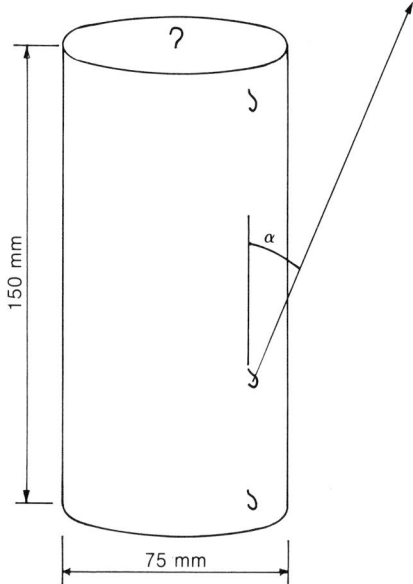

Fig. 1. Model suction-emplaced pile in sand

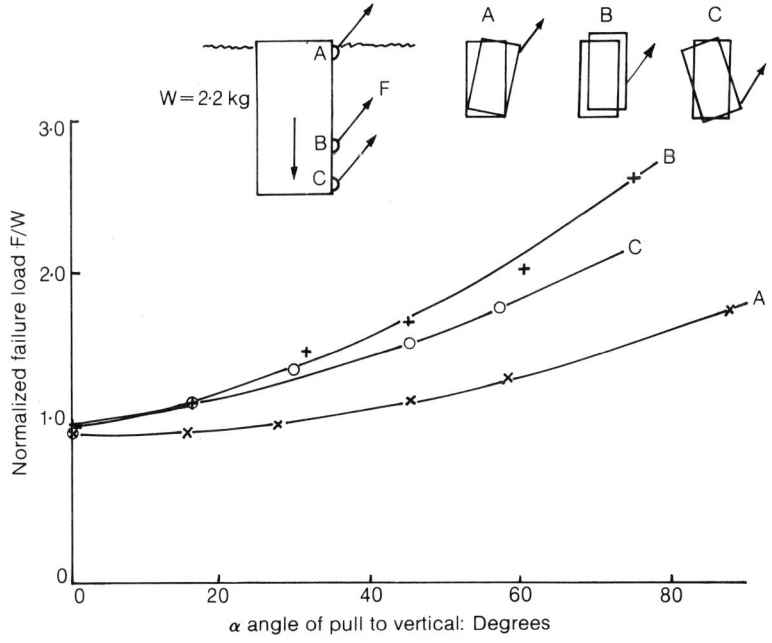

Fig. 2. Dependence of pile capacity on mode of failure

necessary periodically to reapply suction in order to ensure
that the cylinder does not work its way out of the ground?
It is hoped that initial tests will be followed in 1982/3 by
a more extensive series of geotechnical centrifuge model
tests to study this problem, and the Cambridge Soil Mechanics
Group would like to hear of any other studies that have been
made.

MR Q. WILSON, *Robert Gordon's Institute of Technology (RGIT)*

9. Marine suction anchors are a family of devices which
modify the usual hydrostatic pressure gradient in order to
produce effects which increase the holding or force-resisting
capacity of the device. Many variations on this theme are
possible, but three main types of anchor unit can be recognized.

10. First, the inverted cup (Fig. 4). The cup skirts may be
embedded by pushing the anchor downwards, or by a combination
of anchor weight and applied suction, or by a combination of
anchor weight and soil fluidization produced by integral
water jets. Cup-type anchors can be used in a variety of
soil types[1]. Suction piles[2] employ active suction only
during the installation process.

11. Secondly, the buried suction anchor (Fig. 5). The first
examples of this type[3] were concrete hemispheres connected
directly to the anchor chain, and having integral ducting and
water jets. Experience during small-scale sea trials (note
the collapsed hoses in Fig. 5) led to the conclusion that
anchors of this type were best used as footings connected
rigidly to a surface frame. This arrangement allows the
anchor group to react to forces in any direction (Fig. 6).
Current work at RGIT is directed towards the investigation
of the behaviour of a group of such footings, but preliminary
investigations suggest that failure under inclined loading
is by toppling of the complete anchor in the direction of
the applied load, with the uplift resistance of the 'rising
footing(s)' being the factor controlling the break-out force,
at least for the geometry shown. Information on the uplift
resistance of isolated buried suction anchors of hemispheri-
cal form buried in sand is given in reference 4.

12. Thirdly, surface-attaching suction anchors. A slab
anchor with the pressure differential maintained by a drain
or reservoir has been proposed by Schofield[5]. I carried out
preliminary trials with a 'conforming suction anchor' during
1981. This latter device (Fig. 7) is a partly porous flexible bag containing coarse sand or gravel. When dropped on

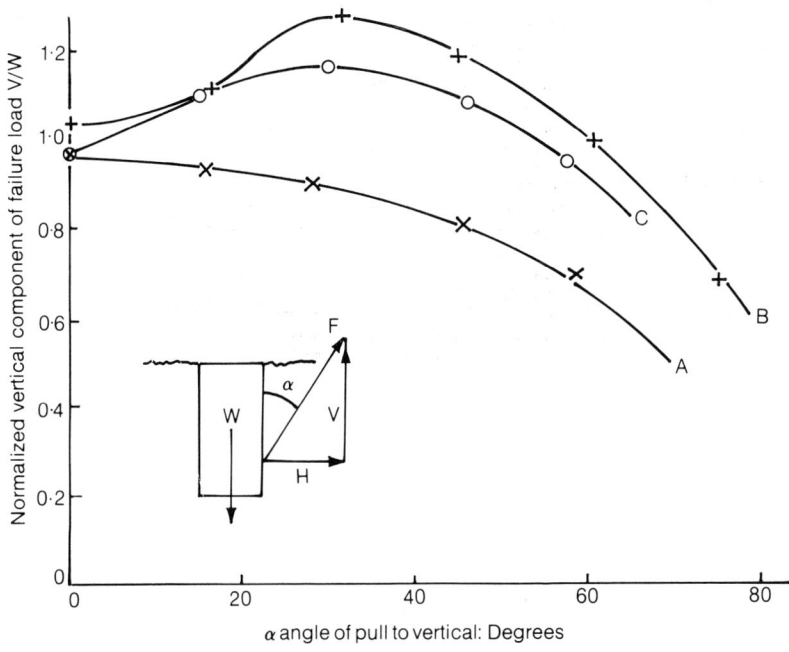

Fig. 3. Maximum vertical capacity of pile

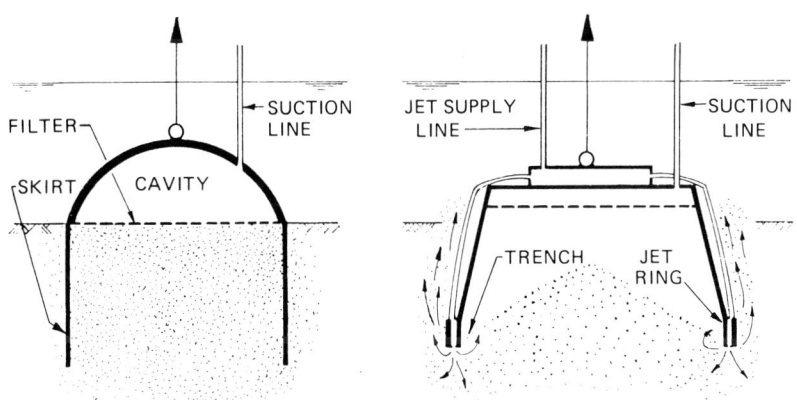

Fig. 4. Inverted cup anchor

Fig. 5. Buried suction anchor

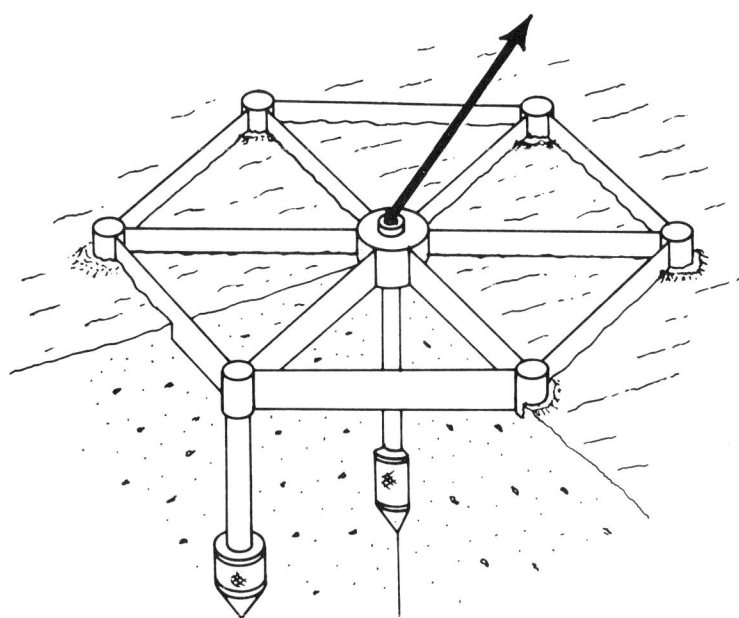

Fig. 6. Buried suction anchors connected rigidly to a surface frame to allow the anchor group to react to forces in any direction

to the sea bottom, it conforms (within limits) to the surface on which it lands. Development problems with the flexible skirt remain, but a mean vertical pull-off force of 28 kN from a clean sandy bottom has been attained. The prototype anchor weighed 4 kN in air and the applied suction pressure was 81 kN/m^2. If, as I believe, the problems with the flexible skirt can be solved, a larger conforming anchor of at least 600 kN (60 t) capacity is feasible. The characteristics of the device would be that it would not penetrate the sea bottom, so that it could attach to shallow soil beds such as sand over rock or to surfaces such as concrete slabs or steel plates. Retrieval would be easy, and since there is no dependence on a hooking or ploughing action the probability of causing damage to sub-sea equipment would be greatly reduced.

13. Although these comments are concerned with particular types of anchor, possibly the most important idea is the active control of soil pore pressure and soil strength.

14. At its most basic level a suction device could be brought into use in order to minimize or prevent pore-pressure build-up during cyclic load application. A modest level of applied suction could create a significant increase in the holding power of a gravity anchor. Higher levels of suction (of the order of 100-1000 kN/m^2) take us into the realms of true suction anchors which can be highly efficient, but which depend on the maintenance of the pressure differential for a large proportion of their break-out resistance. A system such as Schofield's drain or reservoir, combined with some sealing of the sea bed by impermeable sheets, is one way of obtaining an increase in the holding reliability of the system.

15. At the present stage in suction anchor development, and with the exception of the suction pile, the most likely applications are specialist and short term. The ship salvage operation is a possible example. Another is mooring in shifting sands, such as in the area where the Lutine sank off the Dutch coast (treasure hunters please note). Small suction anchors could have a value in providing force reaction points for moving items of sub-sea equipment such as pipes. Larger suction devices might even, one day, be used with economic advantage during the positioning of offshore structures.

MR ALBERTSEN and MR BEARD, *Paper 5*

16. In reply to Mr Dell, the navy Stockless, STATO, and Bruce anchors are depicted in Fig. 4 of the Paper. Stevfix, Stevdig, Stevmud, and Hook anchors, manufactured by Vryhof Ankers BV in Holland, are all similarly configured and are shown in Fig. 8. The Hook differs from the others in that its fluke can act on only one side of the shank. The Bruce twin-shank anchor is very similar to the Bruce anchor shown in Fig. 4 of the Paper except that the shank is constructed of two plates, which reduces shank frontal area and enhances anchor embedment.

17. Anchor lead was measured on the deck of a barge and at the anchor shackle. By knowing the shape of the catenary the load at the sea floor can be calculated. With this data and the anchor resistance, the amount of load carried by chain in contact with or buried in the sea floor is known. Reference 9 gives a full description of the instrumentation system.

18. There is a difference between anchor efficiency and mooring efficiency. The first is the ratio of the anchor's capacity to its weight, the second is the ratio of the total holding capacity of chain-anchor or wire-anchor components to anchor weight. As has been noted, an anchor used with chain may not dig as deeply as an identical anchor used with wire and, hence, the anchor would hold less (assuming the pull at the sea floor is near horizontal in both cases, which requires considerably longer scopes of wire than chain). However, as stated in note (5) of paragraph 24, the overall mooring efficiency would remain about the same. This is because with wire the anchor digs deeper and holds more, but little capacity is added by the wire. On the other hand, with chain, the anchor holds less because of its shallower burial, but significant capacity is added by the chain. This effect has been observed in field test results in a clayey sea floor (reference 9). Recent (unpublished) results of model anchor tests in medium dense sand indicate the opposite: a consistently higher anchor-holding capacity with chain. Apparently the chain disturbs the soil in front of the shank, allowing for deeper penetration than occurs with wire. This behaviour has not been verified by field tests.

19. We agree with Mr Bruce's comment that under some circumstances it is possible, because of the holding capacity produced by the chain, for a deep burial anchor to produce higher total holding capacity at the barge and see less load at the anchor than a shallow burial anchor. However, we do

Fig. 7. Conforming suction anchor

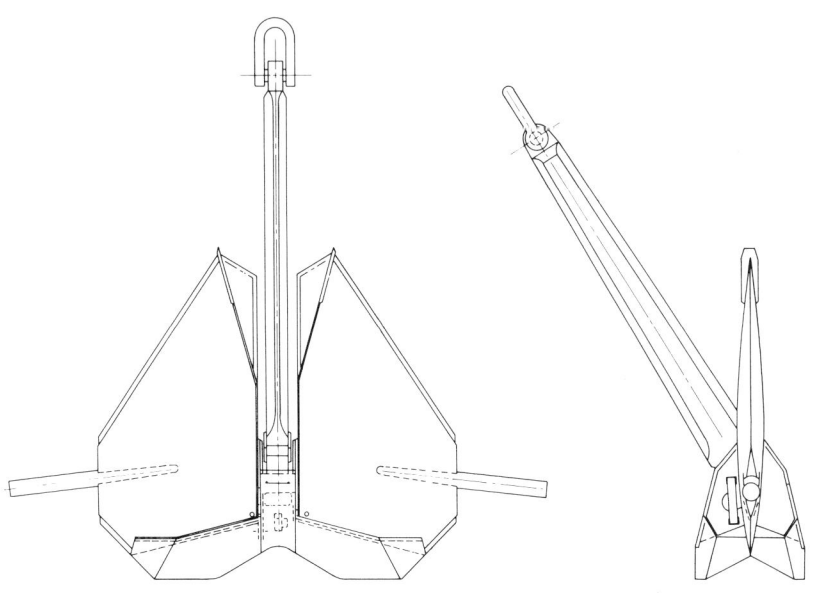

Fig. 8. Typical configuration of a Vryhof anchor

not believe that this is universally true. A variety of soil conditions can exist that preclude deep anchor burial and could lead to the selection of a shallow burial type anchor to satisfy a requirement.

REFERENCES

1. WANG M. C. Vertical breakout behaviour of the hydrostatic anchor. Report CR 74.005, 1974, Civil Engineering Laboratory, Naval Const. Batt. Center, Port Hueneme, Ca, USA.
2. CUCKSON J. The suction pile finds its place. Offshore Engr, 1981, Apr., 80-81.
3. WILSON Q. and SAHOTA B. S. Suction anchors. Paper EUR 16, EUROPEC (conference and exhibition), 1978.
4. WILSON Q. and SAHOTA B. S. Pull-out parameters for buried suction anchors. Paper OTC 3816, Offshore Technology Conference, 1980.
5. SCHOFIELD A. N. Improvements in hydraulic engineering installations, UK Patent 1 367 881, Sept. 1974.